Seeing Sound

SEEING SOUND

WINSTON E. KOCK

The Bendix Corporation
Southfield, Michigan

WILEY-INTERSCIENCE

A Division of John Wiley & Sons, Inc.
New York • London • Sydney • Toronto

To Ralph K. Potter

Preface

Why should we wish to "see" sound? What do we expect to gain by visually portraying a phenomenon that we have always perceived so effectively with our ears? An early maxim states that seeing is believing, and the history of science progress is replete with the efforts of experimenters to reduce the observation of physical happenings and measurements to something that can be seen.

Today visual methods of indicating the important measurements and observations of our everyday life are commonplace. The speedometer in our car, the thermometer at our front door, and the hands of our wristwatch all convert into visible, observable entities the variables that are important to us in our way of living. In the science laboratories of colleges and industry we also find extensive use of visual display, particularly in the instruments that sense changes in certain variables and pass on to our view their measurements through dials, pointers, or moving spots of light. Often the visual presentation helps the experimenter to understand what is taking place; in other cases it permits him to analyze accurately the individual factors involved in the phenomenon or experiment.

It is therefore not surprising to find that attempts were made quite early to study sound phenomena visually. These included the sensitive gas flame (whose height responded to the impinging sound waves) and the "phonautograph," in which a stylus attached to a diaphragm touched a moving smoked surface and traced a curve corresponding to the incident sound. The reasons for these attempts to portray sound phenomena, for their continuation, and for our devoting this little book to a discussion of visible sound are various. Here are a few of them:

1. By delineating the space patterns of sound waves as they propagate, we can observe and measure the effects of the diffraction and refraction processes that sound waves undergo when they strike obstacles or pass through various structures.

2. Because some sounds do not lie within our range of hearing, they require a means of portrayal if their properties are to be fully understood.

3. Some sounds originate in the ocean, and their visual portrayal has proved quite useful in their detection and analysis.

4. We can often understand more about the nature of the complicated sounds of speech and music when the analysis is presented to us in visual, rather than audible, form. (In one instance the last messages of an aircraft pilot about to crash were deciphered by the subsequent examination of a visual analysis of his words as recreated from a tape recording.)

In this volume many kinds of sound will be portrayed and discussed. We shall see natural and man-made sounds that occur in the oceans, we shall see the sounds of speech and music, the sounds of aircraft, and the sounds of noise.

I hope that through visual presentation of various sounds the reader will gain a clearer insight into their nature and that this will further his interest in the many problems that still confront the acoustician.

WINSTON E. KOCK

Southfield, Michigan
June 1971

Acknowledgments

I am indebted to F. K. Harvey of the Bell Telephone Laboratories, who produced most of the photographs of the sound-wave space patterns of Chapters II, III, and VII, to R. K. Potter, B. P. Bogert, K. H. Davis, Homer Dudley, H. K. Dunn, R. L. Miller, W. R. Bennett, and J. L. Flanagan, all of Bell Telephone Laboratories, for the sound spectograms of Chapters IV, V, VI, and VII, and to R. K. Mueller of the Bendix Research Laboratories and Lowell Rosen of the National Aeronautics and Space Administration for the holograms of Chapter VII.

W. E. K.

Contents

Seeing Sound

I   *The Nature of Sound*

Sound is created whenever something moves: the stretched head of a drum, the pounding waves of the sea, the wind moving through the trees. But just as all of these motions are different, so is the sound they create. The nature of any particular sound depends strongly on the motion that creates it. If the motion is an extremely rapid to-and-fro one, the sound will comprise vibrations that are rapid, or high pitched. A less rapid to-and-fro motion will produce a sound of lower pitch, or frequency. The rapidity of the repeated motion has been usually specified as vibrations, or cycles, per second, but recently the term "hertz" (after the German scientist Heinrich Hertz) has become increasingly used.

Thus we see that sound has at least one property that permits it to be distinguished from another: its pitch, or frequency. Because the pitch range of our ears is limited, we can only hear sounds whose frequency lies above about 15 hertz (or cycles per second) and below 15,000 or 20,000 hertz. Hence, if the vibratory motion is too slow (below 15 hertz) or too rapid (above 20,000 hertz), the sound created will not be audible to us. We do, however, have instruments that can detect sounds whose frequencies extend beyond these limits.

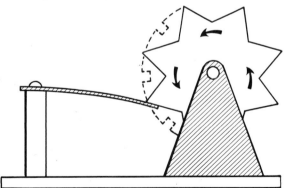

Figure I-1. A rapidly rotating toothed wheel causes a card to generate sound. A wheel with shallow grooves (shown by the broken line) generates sound of lesser loudness.

Just as the rapidity of the to-and-fro motion determines the pitch, the magnitude of the motion determines the loudness. A card held against a rotating toothed wheel, as in Fig. I–1, produces a high-pitched note when the wheel is turning at high speed and a note of lower pitch when the wheel turns more slowly. Likewise, if the teeth of the wheel are shallow (as indicated by the broken line), the generated sound will be less loud than the sound produced by a wheel with deep teeth and a more vigorous motion of the card. We see that another distinguishing property of a sound is its loudness (its *intensity*).

Still another property is quality, which can range from extreme complexity to extreme simplicity. The tone created by a struck tuning fork mounted on its resonator box is an example of a sound that is considered to be very simple in quality, only one frequency, or pitch, being involved. The tone can be loud or soft, but it possesses only a single pitch, which depends solely on the structure of the tuning fork.

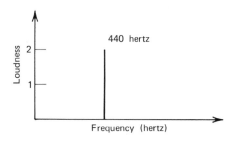

Figure I-2.   A sound wave comprising but one single frequency.

Figure I–2 indicates one way of graphically representing this extremely simple tone. The coordinates are frequency and loudness, and we have indicated that the pitch, or frequency, of the single note is 440 hertz* and its loudness is 2 "units."

If the tuning fork were smaller, it would produce a similarly simple tone on being struck, but its pitch would be higher. With two forks of different size we can create a more complex sound by striking both of them at once. The sound we then hear consists of two frequencies rather than one. Figure I–3 shows the representation of this sound, consisting of the tone of the first tuning fork (frequency 440 hertz, loudness 2 units) combined with the softer tone of the smaller tuning fork (frequency 880 hertz, loudness 1 unit).

We can create sounds that are much more complex than this two-tone

*This is the frequency of the note designated as "A above middle C," the note to which symphony orchestra musicians tune their instruments.

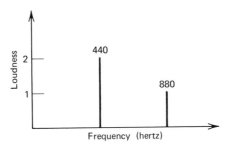

Figure I-3. Two single-frequency sounds.

note; for example, an organ chord constitutes a sound that is made up of a multitude of frequencies. Even the twang of a single mandolin string is quite complex. This is because a stretched string can vibrate not only in its fundamental mode (as the tuning fork does) but also in many other modes. These modes of vibration generate additional tones, whose frequencies happen to be multiples, or harmonics, of the fundamental frequency of the string tone. The various harmonics, or overtones, that are generated add appreciably to the complexity of the tone. The representation of the harmonic series of a tone rich in overtones is shown in Fig. I–4.

However, because they are single tones or combinations of tones, and hence "musical" in nature, all of the sounds we have mentioned so far are not as complex as the sounds of noise. The howling wind of a storm, the rumble of thunder, the noise of a jet aircraft are all sounds whose spectrum of frequencies is very broad. Generally we cannot assign a "pitch" to these noisy sounds, although certain regions of frequency may be stronger than others. Their spectrum representation therefore shows no single "lines,"

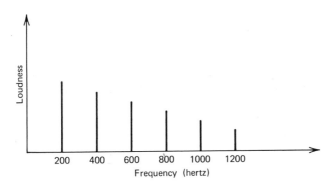

Figure I-4. A periodic wave may have many harmonics, all of which are integral multiples of the fundamental frequency.

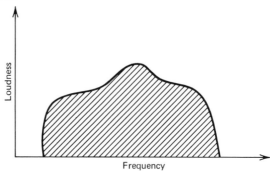

Figure I-5. Noiselike sounds are not periodic, and the spectrum of such a sound extends over a wide frequency band.

as in Figs. I–2, I–3, and I–4; we find rather that broad areas of frequency predominate, as indicated in Fig. I–5.

Recapitulating, we can state that a sound has loudness, a certain quality, or complexity, and perhaps pitch, depending on its frequency content. Some sounds, such as those involved in music, have a frequency structure wherein the overtones are all harmonic to a given fundamental tone; such sounds are said to possess pitch (generally identical with the frequency of the fundamental tone). Others have no nice, exact relationship between the frequencies that constitute the entire sound; they are noiselike in quality.

Let us now examine what happens after a sound has been generated. How does the sound, or "disturbance," travel? How does it propagate, for example, to our ears?

First of all, for airborne sounds the disturbance involves a physical movement of the air itself. We noted that movement creates sound, but for its propagation sound requires a medium in which the disturbance can move out and away from the generating point. Water waves cannot exist without water, and airborne sounds cannot propagate without air. The sound of an electric doorbell becomes very weak if the bell is placed in a bell jar from which the air is evacuated. Of course sound *can* propagate in media other than air. It can propagate in water, in other liquids, and in solids. Because sound does travel in solid materials, the evacuated-bell-jar experiment just mentioned was historically not very convincing. Several experimenters who tried it did not cushion the bell properly, and the sound was heard even when the air was evacuated because of transmission through the supporting structure.*

*Athanasius Kircher and Otto von Güricke both made the bell-jar-in-vacuum experiment around 1650 and concluded that air was not necessary for the transmission of sound. Robert Boyle repeated the experiment in 1660 with convincing proof that air *is* necessary.

Assuming then that a medium is present, what is the mechanism of propagation? In the case of water waves we can see that a disturbance, caused on a still pond by a stone thrown into it, propagates outward in the form of ever widening rings of ripples, or wavelets. We can also envision that a wave generator in the center of the pond, producing a rhythmic up-and-down motion of the water, would create steady, continuous waves, constantly moving out from the source. We shall see in the next chapter (in photographs of the space patterns of sound waves) that sound too propagates as waves.

Because our atmosphere is three-dimensional, unhindered sound waves spread out in not only two directions (as water waves on the surface of a pond) but in three, as ever increasing spheres. Because the fixed amount of energy involved in the original sound must be spread out over an ever larger spherical surface, the loudness, or strength, of the sound grows weaker and weaker as the distance between the sound source and the listener increases.

In watching water waves in a pond we note that they travel out with rather constant speed. Sound waves too have a speed with which they propagate. When a thunderstorm is nearby, we observe that the lightning flashes are followed almost immediately by the thunderclaps, whereas when the storm is more distant, the lightning may precede the thunder by many seconds. This is because the lightning arrives at our eyes almost immediately, whereas sound travels much more slowly. The time lag, or delay, between seeing the lightning and hearing the thunder occurs because sound takes much longer to reach us than does the light of the lightning flash. Sound travels at a speed of about 1100 feet per second (at sea level), whereas light waves move at almost 186,000 miles per second—almost a million times faster. If a thunderclap follows the lightning flash by more than 5 seconds, we can reckon that the stroke of lightning took place at least a mile away.

Taken together, the velocity of sound and the frequency of a tone determine a property referred to as wavelength. On observing water waves on a pond, we note that the circular ripples are sometimes small and sometimes large; that is, the spaces between the wave crests or between wave troughs can be different, depending on the size of the object that caused the disturbance. If as these waves approach us we hold our fingers so that they almost touch the surface of the water, we can feel each wave crest as it passes our fingers. If they are widely separated, the successive wave crests will touch our fingers less frequently as they pass by than wave crests that are close together. We would say that the crests of waves whose crest-to-crest separation was large touched our fingers with lower frequency than did the crests of waves for which this separation was

short. Thus in the case of water waves at least we see that for a given velocity of wave propagation waves that have small spacings between crests
(i.e., short wavelengths) have a higher "frequency," and those with *long*
wavelengths have a low frequency.

A similar thing happens with sound waves. At the point where the sound
is received or detected, its presence is made evident by variations in air
pressure—the same variations that are produced at the sound source by
the movement of the sound-generating surface. We hear sound because
the varying pressure of sound waves pushes our eardrums in and out.
Each time it reaches us, a wave crest, or high-pressure area, pushes our
eardrums inward. If new wave crests follow in rapid succession, our eardrums move rapidly back and forth—that is, with a high "frequency." If,
on the other hand, the wave crests (high-pressure points) are far apart,
then, because of the fixed velocity with which the sound-wave crests
propagate, our eardrums will experience a slower vibratory motion—that
is, one of lower frequency.

The velocity of sound in air is fairly constant. As a consequence the
frequency of a sound is directly related to the distance between the wave
crests it creates; that is, the frequency, or pitch, of a tone has a definite
relationship to the wavelength of the sound wave. We can express this
relationship as follows: the wavelength equals the velocity divided by the
frequency. Stated in another way, the wavelength is inversely propor-
tioned to the frequency, the constant of proportionality being the sound
velocity. In the next chapter we shall see from Fig. II–11 that the properties and propagation characteristics of sound waves are very similar to
those of water waves. That figure illustrates the concept of wavelength
just described—namely, that the wavelength corresponds to the radial
distance between any two of the wave crests shown.

II   *Seeing Sound Waves*

To make a phenomenon or a measurement visible we need a device that can sense the physical aspects of interest in the phenomenon and convert them into a visual pattern. In a thermometer changes in temperature are sensed by the inherent expansion and contraction with temperature of mercury or alcohol and displayed by placing the thermometer tube against a white background. To portray visibly the speed of an automobile we use a device that senses the speed of the wheels' rotation and then adjusts correspondingly the position of a pointer on the speedometer.

To make sound visible we use an electronic device called a microphone. This device senses the varying sound pressure of the passing sound wave and converts it into a varying electrical current. Thus the microphone in a telephone converts voice sounds to electrical signals and transmits them to the other end of the telephone line, where they are reconverted into the original spoken sounds by means of the telephone receiver. By means of the microphone, a loud sound is transformed into a large electrical signal and a weak sound into a small signal.

The electrical signal of the microphone can be made visible by causing it to light an electric light bulb. A loud sound (strong electrical signal) produces a bright light, and, conversely, a weak sound produces a dim light. Because it cannot respond rapidly, an ordinary light bulb cannot follow the rapid variations in the loudness of the sound waves. A small neon lamp, however, can, and so in the display method to be described a neon bulb is employed. In a noisy area the neon lamp will be brightly lit, whereas in a quiet location it will be dim or not lit at all.

To measure the sound intensity at all points in the area and make a picture of this intensity pattern the neon light is affixed directly to the microphone itself, and the variations in its brightness are recorded photographically as the microphone—lamp combination is moved about in the area of interest. The brightness of the lamp at a particular spot is then indicative of the loudness of sound at that spot. To record this brightness pattern photographically a camera is set at a time exposure and aimed at the area of interest. Then, as the microphone—light combination scans the area,

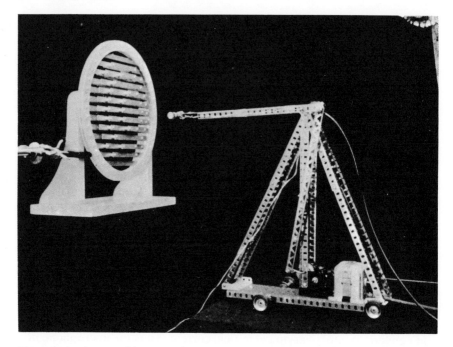

Figure II-1. A motor-driven scanning device of the walking-beam variety made from a toy construction set.

the camera records the light-intensity variations from spot to spot. The result is a photographic record of the loudness pattern of sound in a particular area.

Figure II-1 shows such a device for scanning a sound field with a microphone–light combination made from a toy mechanical construction set. As the arm moves up and down, the microphone and the light at its extremity move in a large circular arc. At the same time the entire device is caused to roll slowly to the right. In the process a zigzag scanning pattern of the microphone and light is created.

Figure II–2 shows the result of a time exposure of the light from this device after it has traversed a steady sound field existing in front of an acoustic lens, a device that focuses sound waves in the same way that a glass burning lens focuses and concentrates sunlight.

In this picture an ordinary (incandescent) electric lamp was used instead of a neon lamp, and the delay between a strong signal and a strong light causes the upstrokes to be portrayed too high and the downstrokes too low. Also, far too lapid a horizontal motion was applied to the carriage, causing the up-and-down strokes to be too widely separated.

Figure II-2. When an ordinary incandescent lamp is used in the scanner of Fig. II-1, an undesirable thermal lag results. Also, the strokes are too far apart.

Figure II-3. When the rapidly responding neon tube is used as a light source, the acoustic beam of the acoustic lens becomes discernible, but the scanning device still produces too coarse a pattern.

Figure II–3 is a similar record made by using a neon tube for the lamp. Here the bright portions of the zigzag strokes correspond to the central focused area in which the sound is loud, and the darker ones correspond to areas in which the sound is weaker. A focused sound pattern is recognizable, but, again, the motion of the carriage was too rapid, causing the vertical scan lines to be too widely separated.

To overcome this difficulty and to provide a larger scanned area a sec-

Figure II-4.   An improved model of a scanner.

ond version of the scanning device was built, as shown in Fig. II–4. On the right is the camera, which is set at a time exposure. In the center is the scanning rod; it has a small, black microphone and a neon bulb affixed to its end. The acoustic lens is the white object in the upper left; the black hole behind it is the mouth of a horn from which the sound waves originate. An electric motor produces the up-and-down motion of the microphone and also causes the entire assembly to move slowly toward the reader.

Figure II–5 shows how this version portrays the sound field portrayed in Figs. II–2 and II–3 by the first, simple, mechanism. In this portrayal the vertical scan lines produced by the zigzag motions of the neon tube are sufficiently close together to overlap and thus be no longer individually visible. The bright center of the picture is the area in which the lens has caused a concentration of sound — that is, the area in which the sound is more intense.

Figure II–6 portrays the sound field produced in front of a long horn, or megaphone. Megaphones are used to cause a person's voice to be di-

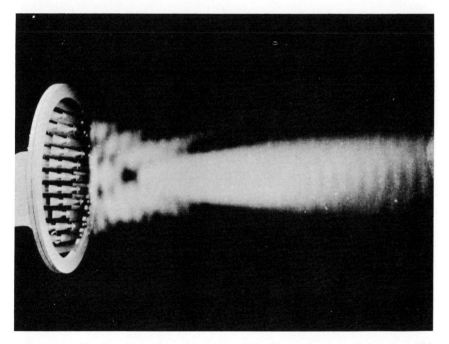

Figure II-5. The finer grained scanning device of Fig. II-4 produces a smooth pattern. This 10-inch acoustic lens is focusing sound waves having a frequency of 9000 hertz and a wavelength of 1.51 inches.

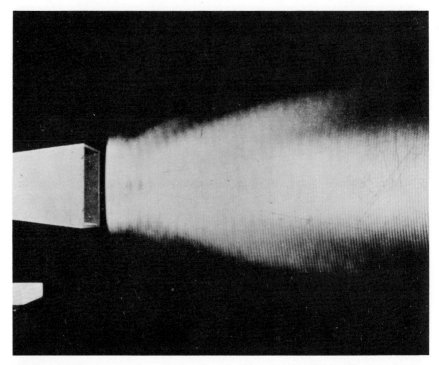

Figure II-6. The pattern of sound intensity caused by the pyramidal horn loudspeaker can be portrayed visually with the apparatus of Fig. II-4. The aperture of the horn is 6 square inches; the sound waves being radiated have a frequency of 9000 hertz.

rected more strongly in a given direction; the photograph indicates why this happens. The brighter areas are indicative of the fact that much stronger sound intensities exist in the direction in which the horn is pointed, the strength of the sound being increased in the forward direction. Thus a horn provides a way of concentrating or directing the sound energy along the axis of the horn.

In addition to making the intensity patterns of sound visible this photographic technique is capable of portraying the progress of the sound waves themselves, much as the progress of water waves can be observed on a still pond.

The photographic portrayal of wave motion makes use of the interference effects caused by two sets of waves. If two pebbles are simultaneously dropped at some distance apart in a pond, each will create its own set of circularly expanding waves. When they meet, the two sets of waves are said to *interfere* with one another. At the points where both sets of

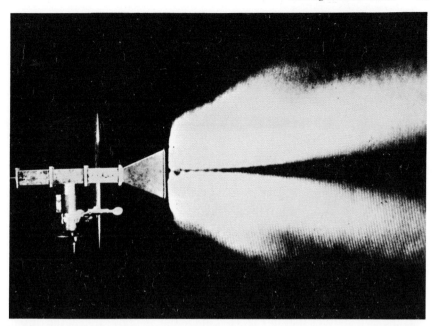

Figure II-7. The pattern of sound generated by two adjacent single-frequency sound sources of opposite polarity.

waves have crests, their combination results in a still greater wave height. Where two troughs coincide, the combined wave trough will be deeper. At points where one wave set has a crest and the other a trough, the two waves act in opposition, the crest of one filling the trough of the other. At these points *destructive interference* takes place. At the points where the wave amplitudes add, *constructive interference* takes place.

The interference effects observed in water waves also occur in sound waves. If the sound waves that issue from, say, a horn are created by two sound generators acting in opposition to each other, one of these sources is creating a crest, or high-intensity wave, while the second source is simultaneously creating a trough, or low-intensity portion. Sound sources that act in opposition to each other are said to be *out of phase*.

Figure II–7 shows the pattern created by two opposing sound sources placed one above the other. Along the line equidistant between the two sources the waves have canceled each other and there is no sound. A black area hence appears along the horizontal center line. All along this line the two sources create destructive interference. Whereas in Fig. II–6 (where there was only one source of sound in the horn throat) a full white area appears in front of the horn aperture, in Fig. II–7 the white central area is divided by a black area along the central line.

Figure II-8. The pattern of sound generated by two separate single-frequency sound sources of opposite polarity.

If the two sources that in Fig. II–7 are acting in opposition at the throat of the horn are separated, the pattern of the generated sound will appear as shown in Fig. II–8. Here the two sources are now at the orifices of the two pipes shown at the left. Again, along a horizontal line midway between the two sources there is a black area, caused by the destructive-interference effect of the two sources. In addition there are other dark areas, indicating that the sound is weak there. This occurs because the sources are now three wavelengths apart, and hence the new pattern has several areas of destructive interference in which the wave crests from one source combine with the troughs from the other.

Figure II–9 indicates the manner in which two waves, A and B, can add either constructively, as shown at the left, or destructively, as shown at the right. In the first situation the wave crests and troughs are increased

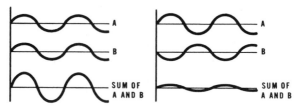

Figure II-9. Two waves of the same wavelength add (at the left) if they are in phase and subtract (at the right) if they are out of phase.

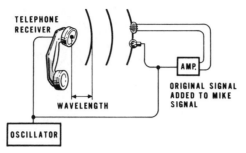

TELEPHONE
RECEIVER

AMP.

ORIGINAL SIGNAL
ADDED TO MIKE
SIGNAL

WAVELENGTH

OSCILLATOR

Figure II-10. Combining the microphone signal with the oscillator signal results in wave addition or wave subtraction, depending on the position of the microphone relative to the sound source.

in size; in the second situation they are decreased. If, in the second case, the two waves are of exactly the same magnitude and in exact opposition, complete cancellation occurs and no waves at all remain. Along the center line of Fig. II–8 the wave amplitudes are exactly alike and complete cancellation occurs, but because the other black areas are nearer to either the upper or the lower sound source, the signals in those areas are not equal in strength, and accordingly complete cancellation does not occur.

The fact that cancellation effects occur at wavelength intervals permits us to exploit these interference effects for portraying wave motion. To do this a second nearly identical electrical signal is added to the signal picked up by the microphone. In Fig. II–10 a signal generated by an electrical oscillator is added to the amplified microphone signal. This combined signal is used to affect the brightness of the neon lamp. The field being investigated is that produced by the sound issuing from a telephone receiver energized by single-frequency waves coming from the oscillator. These single-tone sound waves, like water waves, generate a moving pattern of circular waves centered at the receiver.

In the figure the wave crests are indicated by solid black, circular sections, with the troughs located between them. The microphone is shown as being (momentarily) at the point of a wave crest, and it is further assumed that at that instant the output of the oscillator is at a wave crest. Since the electrical outputs of the two sources (microphone and oscillator) are added, higher electrical crests will result, and instants later, when a wave trough moves into the position of the wave crest (again for both sources), the combined electrical trough will be lower, as was seen at the left of Fig. II–9. Constructive interference will result, and the neon tube will be brightly lit. When, however, the microphone is moved to a position halfway between the crests (the black circular lines), the two electric outputs oppose each other, the signals cancel, and the very small (combined) electrical output causes the neon tube to be dim or extinguished.

By making the electrical outputs from the two sources almost identical, strong interference effects can be made to occur. Thus there will be rings in the space in front of the receiver where the combined signal is very small and other locations (the black circular sections in Fig. II–10) where the crests and troughs again occur simultaneously for the two sources, with the neon light again being bright. At points in space that are one, two, or three wavelengths (or any integral number of wavelengths) distant from the receiver the two signals will add and the neon lamp wil be bright. At the half-wavelength points the signals will subtract and the light bulb will be dim.

As the microphone — neon light combination moves outward from the sound source, the light pattern will match the position of the wave troughs and wave crests of the sound waves emanating from the sound source,

Figure II-11.   When the wave addition and subtraction technique of Fig. II-10 is employed, the actual wave fronts of sound waves can be portrayed. At a frequency of 4000 hertz a telephone receiver is relatively nondirectional, and the sound waves spread out in all directions.

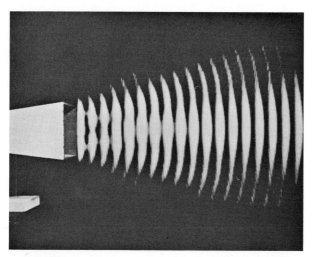

Figure II-12. The wave fronts of the pattern in Fig. II-6.

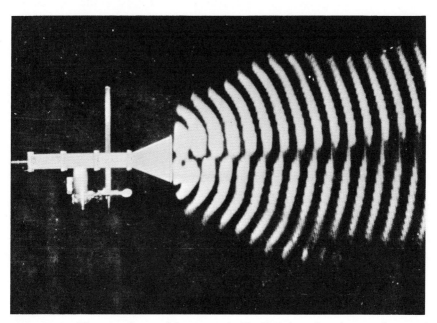

Figure II-13. The wave fronts of the pattern in Fig. II-7 show the existence of reversed polarity; the dark wave troughs at the top are opposite the light wave crests at the bottom.

17

since this light pattern of crests and troughs, as traced by the moving microphone and neon light, remains fixed in space. Because of this, the microphone–light combination can be moved about in any pattern, up or down or sideways, and the light pattern will always repeat itself. Such a pattern of the sound field produced by the telephone receiver is shown in Fig. II–11.

In Fig. II–11 the waves are ever increasing rings whose center is the telephone receiver. The resemblance to water waves on the surface of a pond is evident. In this photograph the telephone receiver is reproducing a pure tone signal of 4000 hertz. Because the receiver is small in terms of the sound wavelength and hence not directional like the horn of Fig. II–6, the waves spread out in a uniform circular pattern in all directions.

In Fig. II–12 the sound pattern of Fig. II–6 is examined by using the wave-front technique just described. The wave fronts here have only a negligible curvature; their thinning out (and disappearance) at the top and bottom is indicative of the very low sound intensity outside the beam produced by the horn.

Figure II–13 shows the wave-front pattern of Fig. II–7. The phase opposition between the top and bottom sources is here made even more evident; the white striations at the top line up in the center with the black striations at the bottom. In other words, a wave crest in the top section meets a wave trough in bottom section.

In the next chapter the procedure just described for examining the patterns of sound waves as they propagate will be used to portray numerous patterns of various types. We shall note how visual portrayal can help us to understand more clearly certain sound diffraction and refraction processes.

III   Some Wave Patterns

In this chapter we discuss examples of sound-wave patterns that can give an insight into such wave phenomena as interference, diffraction, and refraction. Interference and diffraction are the basic processes involved in holography, an interesting optical technique that produces very striking three-dimensional photographs. A more recent development is acoustic holography, which provides a visual reconstruction of sound-wave "photographs." This subject is discussed at length in Chapter VII.

Diffraction effects occur whenever waves strike opaque objects, and wave theory predicts for certain situations some rather unusual phenomena. Some of these predictions are difficult to accept even though the theory itself is quite straightforward.

One such phenomenon is the presence of a bright spot in the shadow of an opaque disk on which light waves are falling. Our everyday observations of shadows cast by opaque objects make it difficult to believe that the intensity, or brightness, of a wave in the deep shadow of a disk is, as Lord Rayleigh stated in his *Theory of Sound,* "the same as if no obstacle at all were interposed." It is not easy to demonstrate this effect for light waves with, for example, a coin, because very short wavelengths are involved, but in his *Theory of Sound* Lord Rayleigh has described methods for demonstrating the existence of the bright spot with high-frequency sound waves, using either the then available sound-detection device, a sensitive gas flame, or "the ear furnished with a rubber tube." He noted that this "region of no sensible shadow, though not confined to a mathematical point upon the axis, is of small dimensions." He further noted that "immediately surrounding the central spot there is a ring of almost complete silence, and beyond that again a moderate revival of effect."

A visual representation of the sound-wave pattern existing in the shadow of an opaque circular disk is shown in Fig. III–1. Made by the amplitude presentation technique described in the preceding chapter, it clearly shows the bright spot, but it still does not demonstrate the reasons for this phenomenon.

The additional information present in a visual representation of the wave

Figure III-1.   The amplitude pattern of sound waves in the shadow of a disk clearly shows a bright major lobe.

motion in the shadow area provides further insight into the causes of the phenomenon. The progress of sound waves in the shadow of a single-edged opaque screen is portrayed in Fig. III–2. The single knife edge acts as a new source, generating cylindrical waves that propagate into the shadow area. In the area at the top the waves are not obstructed by the knife edge, and parallel vertical wave crests continue to progress toward the right. Much weaker cylindrical waves are seen in the shadow area progressing outward from a line located at the knife edge.

Figure III–3 shows the wave pattern of sound behind an opaque circular disk. As in Fig. III–2, circular wave fronts originating at the top and bottom edges are evident in the shadow area. Sound sources thus exist at both the top and the bottom portions of the disk, and the two sets of waves can produce interference effects in the shadow area. Indeed, similar sources of sound exist along the entire perimeter of the disk, and because all of them are equidistant from the central axis, it is understandable that a concentration of energy will be obtained along the axis, the bright spot of Fig. III–1. The central cone of bright and dark spots is surrounded by another series of bright and dark areas. The scanning process permits only the top and bottom areas of this second conical volume to be shown in this photograph, but circular symmetry predicates the existence of a complete annular conical volume of sound energy. The phase of these waves is reversed from the phase of the waves in the central cone; the positive crests (white areas) of this set line up with the negative troughs (dark areas) of the central set.

In order to pass from a positive phase condition to a negative one the sound field between these two regions must go through a zero condition. Figure III–3 portrays this zero-sound area as two black lines with no visible white areas, located immediately above and below the central cone of striations. This cone of silence corresponds to Lord Rayleigh's ring of almost complete silence surrounding the central spot, and the negative-phase areas represent what he called "a moderate revival of effect."

Another wave-propagation process that is made more understandable by visual observation of wave progress is *wave refraction*. One of the

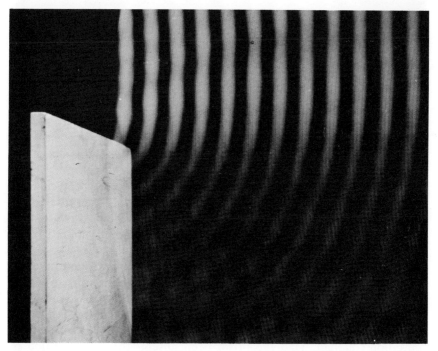

Figure III-2.   Plane sound waves arriving from the left proceed unhindered at the top. In the shadow region below circular wave fronts are seen, caused by diffraction at the edge of the shadowing object.

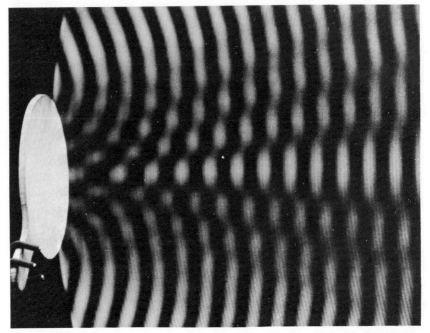

Figure III-3. Sound waves diffracted around a circular disk combine in the shadow region and produce a central beam of parallel wave fronts. This is the bright lobe of Fig. III-1.

simplest forms of refraction occurs in optics when a piece of glass of triangular cross section (i.e., a prism) is made to cause plane-parallel waves to be refracted. The parallel wave fronts originally proceeding in one direction emerge from the prism proceeding in another direction. Since their velocity of propagation is lower in glass than it is in air, light waves passing through the thick portion of the triangular section are delayed more than those passing through the thin portion. An analog often used to explain this refraction process is a row of marching soldiers. If the soldiers at one end of the row march more slowly, their forward motion will be altered in direction.

Figure III–4 shows the refraction of sound waves by an acoustic prism. This prism consists of spaced metal strips, which decrease the velocity of sound waves passing through them. The source of sound is the small horn at the left. An acoustic lens, also made of metal strips, causes a concentration of sound, just as an optical lens causes a focusing, or concentration, of light. Had the prism been absent, the concentration of sound would have occurred in a horizontal direction; the prism causes the beam of the lens to be tilted downward. By portraying the wave fronts

Figure III-4. The focused beam of sound waves formed by an acoustic lens is deflected downward by an acoustic prism.

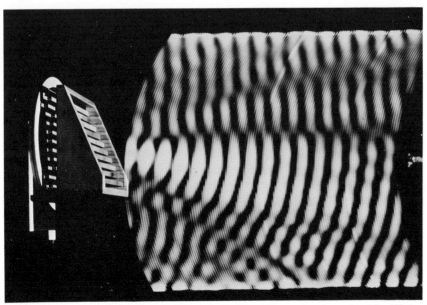

Figure III-5. The wave pattern of the refracted sound energy portrayed in Fig. III-4.

23

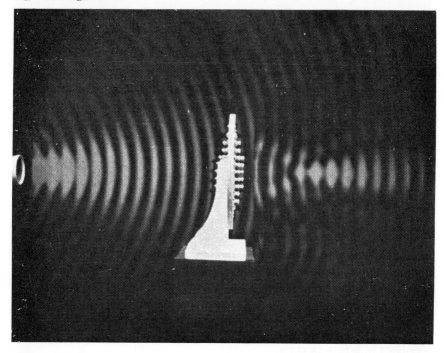

Figure III-6. For an acoustic lens in which the phase velocity is less than free-space velocity, a convex shape causes the phase of waves passing through the lens to be retarded.

of the sound waves leaving the prism, as in Fig. III–5, this tilting of the direction of the wave fronts is made evident.

The simple prism-refraction process can be used to explain the focusing of a lens, which is also a refraction process: the lens redirects the course of wave fronts by changing their velocity. Figure III–6 shows circular sound waves issuing from a horn. On passing through the lens (the acoustic lens of Fig. III–4) they are redirected and curved inward toward a focused area. The lens is thick at its center and thin at its perimeter. In the vertical cross section the top half can be looked upon as a prism (like the one in Fig. III–5) directing the sound downward toward the focal area, and the lower half acts like a prism directing the sound upward toward the focal area. This redirecting effect is clearly seen in Fig. III–6: at the left of the lens the wave fronts are convex, expanding outward, whereas to the right of the lens they are concave, causing energy to be concentrated at the focal point. At the top, where the lens has no effect, the waves continue to expand outward toward the right.

The refractive properties of the metal-strip lens and the prism of Figs.

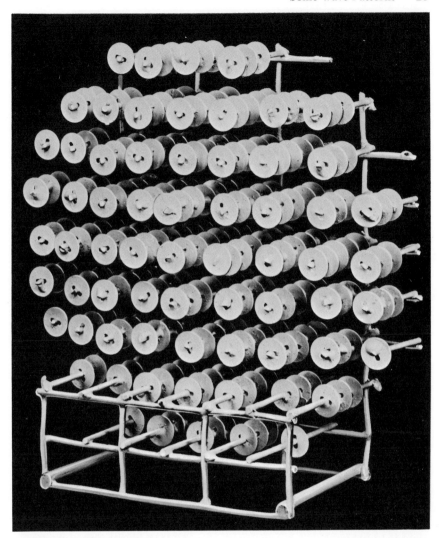

Figure III-7. A structure consisting of rows of disks and having an overall convex shape can act as a lens for sound waves.

III–4, III–5, and III–6 are due to the delaying effect of the elements that decrease the wave velocity.

A similar decrease in wave velocity is produced by an array of disks. An acoustic lens of this type is shown in Fig. III–7. Its convex cross section again causes a concentration of sound waves, as shown in Fig. III–8.

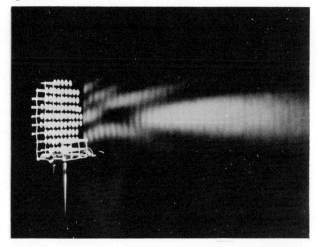

Figure III-8.   Sound waves approaching the lens of Fig. III-7 from the left are seen to be concentrated, or focused, into the white area at the right.

Each row of disks in the lens of Fig. III–7 is equivalent, for sound waves, to a rodlike portion of glass (or dielectric rod) in a lens designed to focus light or electromagnetic waves. Dielectric rods are known to produce an effect for the very-short-wavelength electromagnetic waves called microwaves. They can cause microwaves to be aimed in a given direction, and when made quite long, they can guide such waves from one point to another. There is a theory that the rods and cones of the hu-

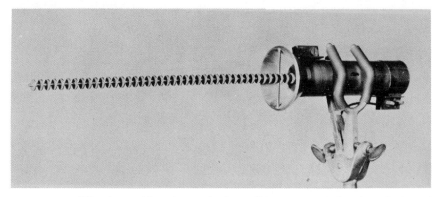

Figure III-9.   When inserted into the mouth of a small horn, a row of disks from the lens of Fig. III-7 can act as a directional "end-fire" radiator for sound waves.

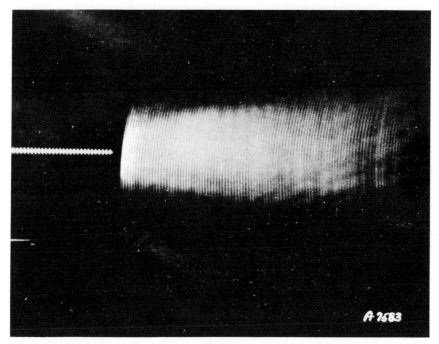

Figure III-10. Sound waves are collimated by the radiator of Fig. III-9.

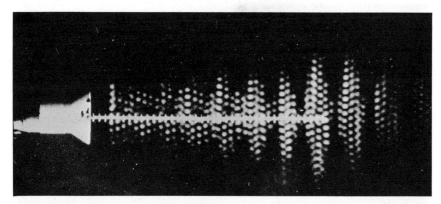

Figure III-11. Waves emerging from the horn at the left would normally possess circular wave fronts. The row of disks causes the velocity of waves near it to be smaller than free-space velocity, and the curved wave fronts are thus converted to plane wave fronts. The directivity of a large-aperture radiator is thus achieved by the end-fire process.

27

man eye cause the light waves entering our eyes to be received in similar special ways because of the rod structure.

The acoustic disk "rods" constituting the lens of Fig. III–7 exert a similar effect on sound waves. A disk rod mounted in the mouth of a horn radiator (Fig. III–9) causes sound waves to be formed into a directional pattern on leaving the end of the rod (Fig. III–10).

The wave pattern shown in Fig. III–11 indicates why this directional effect occurs. Without the rod, the waves emerging from the horn at the left would, as shown in Fig. III–6, possess curved wave fronts, and little directional effect would be observed. The disk rod causes a reduction of wave velocity in its vicinity, as it does in the lens of Fig. III–7, and the wave fronts are therefore flat rather than curved. The radiated energy is thus concentrated in the horizontal direction.

Figure III–12 shows the amplitude distribution in the vicinity of a long, extended disk rod. At the left there is a steady-state propagation condition, and at the right the discontinuity resulting from the rod's being terminated causes reflections that alter the uniformity of the energy distribution.

When the end of a similar extended disk rod is bent into a curved section, as in Fig. III–13, most of the sound-wave energy follows the curved section of the "wave guide" and emerges at the end, as shown by the bright area at the right. The white area at the lower right-hand corner shows that some sound "jumped the track" and remained directed along the line extended from the original long, straight rod section off to the left of the photograph.

Figure II–8 showed the pattern of sound generated by two spaced generators acting in opposition; the horizontal line midway between the sources is one of zero intensity. Figure III–14 shows a similar pattern

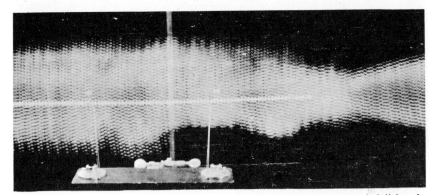

Figure III-12. The amplitude distribution in the vicinity of a long, extended disk rod.

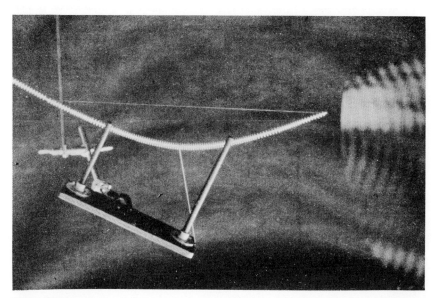

Figure III-13. Most of the sound energy guided by the curved disk transmission line radiates off the end.

Figure III-14. Two separate in-phase sound sources act like two optical slits in creating a diffraction pattern by constructive and destructive interference.

29

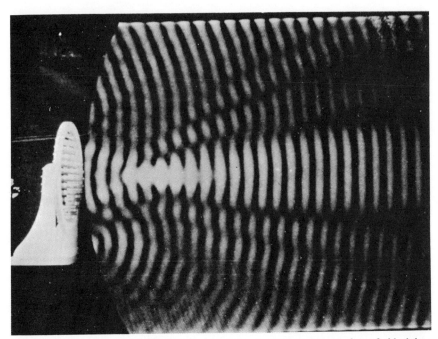

Figure III-15. When high-intensity sound waves are employed, a number of side lobes become evident. In this photograph the phase of the wave fronts is portrayed. It indicates that phase reversals occur between the main lobe and successive side lobes.

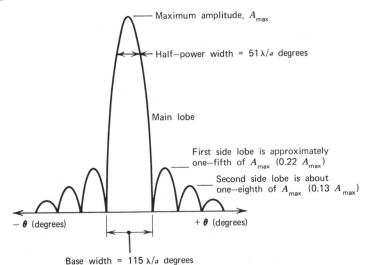

Maximum amplitude, A_{max}

Half—power width = 51 λ/a degrees

Main lobe

First side lobe is approximately one—fifth of A_{max} (0.22 A_{max})

Second side lobe is about one—eighth of A_{max} (0.13 A_{max})

$- \theta$ (degrees) $+ \theta$ (degrees)

Base width = 115 λ/a degrees

Figure III-16. The calculated pattern of a uniformly excited slit aperture, showing the actual magnitudes of the beam width and minor lobes (λ = wavelength, a = aperture).

30

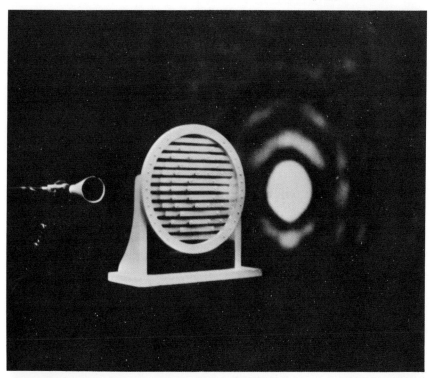

Figure III-17. By sampling the sound field in a plane perpendicular to the direction of the waves, a cross section of the main lobe and side lobes is portrayed. The side lobes are seen to be annular cones of sound surrounding the solid sound cone of the main lobe.

generated by two sound sources acting together instead of in opposition. Here a bright area extends along the line midway between them, indicating the existence of constructive interference along this line, instead of the destructive interference of Fig. II-8.

Two bright areas appear immediately above and below the central white area of Fig. III-14. These three white areas are often referred to as *lobes,* the central one being called the *major lobe* and the remaining ones *minor lobes.* The wave pattern in the shadow of the disk also has lobes. The main lobe is clearly seen in the amplitude pattern of Fig. III-1, and in the wave picture of Fig. III-3 the areas above and below the major lobe are shown to be in an opposing-phase condition. This negative-phase condition—the lining up of the white bars, or crests, with the dark bars, or troughs, in the central area—is typical of radiation lobes. The phases of successive lobes are continually reversed: the central major lobe is usually considered to be positive, those immediately adjacent are negative, the second set of minor lobes is positive, and so forth.

Figure III–15 is a wave-pattern record generated by the same lens that produced the amplitude pattern of Fig. II–7. In this record the wave pattern has been added and the signal strength has been significantly increased to make lower intensity minor lobes sufficiently clear. The first minor lobes (those that flank the major lobe) are again of negative relative phase (their light areas line up with the dark areas of the major lobe), and the succeeding lobes surrounding the first set of minor lobes are of positive phase (i.e., of the same phase as the major lobe).

Figure III–16 is a curve of the theoretical radiation pattern produced by a square aperture. Both the wave amplitude and the phase (flat wave fronts) across the aperture are uniform.

Figure III–16 portrays only a two-dimensional cross section of the three-dimensional sound-wave pattern produced by the lens. The lobes produced by the lens of Fig. III–5 are actually ringlike. This can be shown by causing the scanner to scan a plane perpendicular to the lens axis in-

Figure III-18. The near-field amplitude pattern of sound waves issuing from a 30-wavelength aperture. It is seen that the beam remains collimated for quite some distance in front of the aperture.

stead of, as in Fig. III–15, a plane including the axis. Such a record is shown in Fig. III–17. The major lobe is here seen in cross section to have the expected pencillike form. The surrounding, not quite complete, white ring shows that the structure of the first minor lobe is annular, and the next even less complete ring portrays the next minor lobe. The phase reversal between successive lobes causes annular cones of silence to appear between the white areas.

The portrayal process is also useful in showing how the pattern at an aperture is successively converted to the distant-field radiation pattern as the wave energy moves away from the aperture. We saw in Fig. III–16 the distant-field pattern created when the very-close-in pattern (the amplitude distribution across the aperture) is uniform. But the patterns in between these two end points are also often of interest. Figure III–18 shows the pattern variation that exists very near the aperture of a large plane-wave sound source.

IV   Seeing the Structure of Sound

Most of the sound waves shown so far were single-frequency sound waves and hence quite simple in form (Fig. I–2).

We noted in Chapter I, however, that sound can be very complex. We saw that a sound can have many harmonics (Fig. I–4) or a noiselike quality (Fig. I–5). Finally, the sound itself can change rapidly with time. A visual presentation of these characteristics of sound can accordingly also be useful. Whereas the preceding visualizations showed various *space patterns* of sound, in this chapter we shall describe a way of visually presenting the *structural patterns* of sound. Such a presentation provides a means of visually inspecting the tonal complexity and temporal variations of a particular sound.

We first note that the simple representation of sound used in Chapter I is quite satisfactory for sounds that do not change with time. The steady note of a pipe organ is an example. On being depressed a pipe-organ key generates a sound whose properties remain unchanged until the key is released. The properties of such a sound are hence completely defined by the loudness-versus-frequency representations of Chapter I.

Most sounds, however, vary with time. For example, the tone created by striking a piano key soon dies out, and, because some of the harmonics die out faster than others, the harmonic content of the tone changes. Similarly the sound produced by a trombone changes in pitch as the player moves the trombone slide. The properties of speech sounds vary extensively and rapidly with time. Vowel sounds possess pitch, which can vary while the vowel is being spoken, as can the quality, or harmonic content, of the sound. Speech can change rapidly from a voiced sound (a vowel) to an unvoiced one (a consonant). To represent accurately these various dynamic sounds throughout their complete life history would require a large number of the simple analyses of Chapter I.

We need therefore a different form of portrayal—one that can show changes in quality and pitch as they occur. The most effective process for doing this was developed by Ralph K. Potter and his associates at the Bell Telephone Laboratories. The analyzing device is called a *sound spectro-*

34

graph, and the speech records it generates are called *visible-speech spectrograms.* The main goal of this development was to obtain visual analyses of the sounds of speech to assist in understanding the nature of speech sounds and the problems involved in the telephone transmission of speech. Potter's ingenious technique later proved very useful in many other fields, and the sound spectrograph is now used for the visual portrayal of underwater sounds, aircraft sounds, correlation patterns of sounds, and even signals that correspond to pictures, or likenesses, of persons.

For his presentation Potter selected the passage of time as one coordinate, plotted horizontally, and frequency as the other, plotted vertically. In this form of presentation the numerous harmonics of Fig. I–4 would be shown along the vertical coordinate instead of the horizontal one. The loudness, or intensity, of each frequency component is indicated by the blackness of the marking.

The way in which the analysis is made is sketched schematically in Fig. IV–1. The sound to be analyzed and displayed is first recorded on a magnetic drum, which acts like a loop of magnetic tape of the type used in home tape recorders. Having been recorded, the signal to be analyzed is played back many, many times through a variable-frequency filter by a continual rotation of the magnetic drum. After each rotation, each time the signal is passed through the filter, the frequency of the filter (its position in the frequency range of interest) is changed. The output of this varying-frequency filter is amplified and caused to mark, by means of a stylus, a sheet of electrically sensitive paper wrapped around a second, synchronized, drum. During each revolution the stylus records, by a line of varying blackness on the paper, the varying amplitude of one narrow frequency region of the signal passing through the filter. As the filter moves upward in frequency, the stylus moves with it in synchronism. When the signal from the filter is strong, the stylus marking on the paper

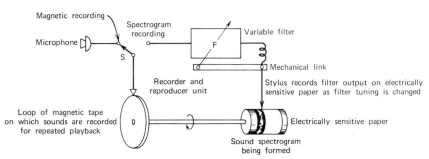

Figure IV-1.   The analyzing procedure involved in the sound spectrograph of R. K. Potter.

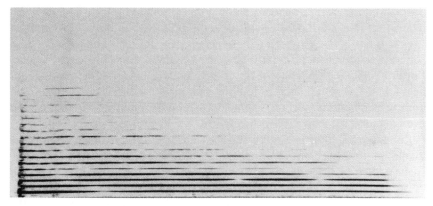

Figure IV-2. Analysis of the sound generated by the striking of a piano key. Here the amplitudes of the fundamental frequency and its harmonics vary with time.

is dark, whereas when the signal is weak, either no mark is printed or only a light-gray mark results.

The sound spectrograph is thus capable of completely analyzing and visually portraying the amplitude and frequency of a sound. Just as a variable-color light filter distinguishes the various colors present in a source of light, so the varying-frequency electronic filter indicates the "colors," or frequencies, present in the original sound. Simultaneously portrayed are the time variations in the strength of each frequency component.

Figure IV–2 is a spectrogram of a sound varying in time. It is the sound produced by striking a piano key. Initially, at the instant of striking (on

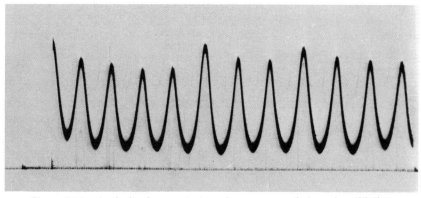

Figure IV-3. Analysis of a sound whose frequency, or pitch, varies with time.

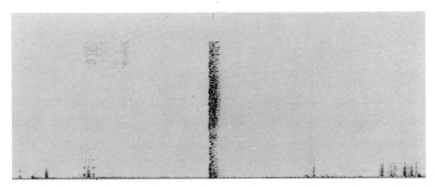

Figure IV-4. Analysis of the noise made by a gun. Inasmuch as this record was made in an anechoic room, the sound, which did not reverberate, was very short in duration.

the left), a large number of harmonics are generated. They are all multiples of the lowest, or fundamental, frequency, the bottom rather dark line. With the passage of time (moving from left to right) the higher harmonics become too weak to be recorded, whereas the fundamental and lower harmonics persist. This is seen at the left of the record, where the markings are very black, indicating that the tones were originally quite loud. As time elapses, the sounds die out, and the marks pass through various shades of gray, becoming continually lighter and lighter. The sound of the hammer blow, at the start of the record, is seen to be noiselike in nature, the gray areas at this point in time extending over the full frequency region.

Another example of a time-varying sound is shown in Fig. IV-3, which portrays the analysis of a tone, without overtones and of constant loudness whose pitch is being varied up and down. It is the analysis of the sound issuing from a loudspeaker connected to an oscillator whose pitch, or frequency, is being raised and lowered. This change in pitch is similar to that produced by a trombone player, who periodically varies the pitch of his instrument by moving the trombone slide back and forth.

The next spectrograms are records of noiselike sounds. A gunshot is a sound comprising a broad spectrum of noise. When it is fired in the open, a gun produces a sound of very short duration. If, however, the gun is fired inside a hall, the sound will usually reverberate for a much longer period. Figure IV-4 portrays a sound comparable to that of a blank cartridge being fired in the open. For this record a toy gun was fired in an anechoic room — a room with walls that strongly absorb sound, thereby preventing its reflection. The spectrogram shows that this sound

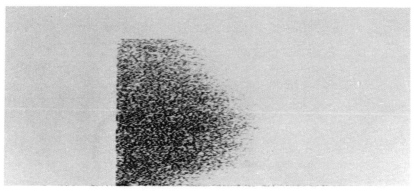

Figure IV-5. The report of Fig. IV-4 sounding in an auditorium. Reverberation extends the duration of the record, the amount of the extension depending on the frequency.

comprises the full range of frequencies (it encompasses the full vertical scale) but lasts for only a short time (it is narrow in width).

Figure IV–5 is the record of the firing of the same gun in an auditorium; here the sound created by the shot reverberates for some time. Although the sound still covers a wide frequency band, some of its areas (i.e., sounds that fall within certain areas of frequency) persist longer than others. These sounds are said to reverberate longer, and the length of time a sound continues to reverberate in an auditorium before dropping one one-thousandth of its original loudness is the auditorium's reverberation time for that sound.

The reverberation time is an important measure of an auditorium's acoustical quality for music and for spoken sounds. Originally it was considered satisfactory to specify the reverberation time for a single frequency, and the standard frequency was set at 512 hertz. (This is approximately the frequency of the piano note C above middle C.) Occasionally this single-frequency specification is still used, and when the term "reverberation time" is employed without specifying the frequency, it is generally understood to mean the reverberation time at the 512-hertz frequency.

However, specifying the reverberation time in terms of one frequency would be truly correct only if the sound-absorption effectiveness of the auditorium walls were uniform over the frequency range of interest. In general this uniformity does not exist. A room with plaster walls could have, for example, at 512 hertz the highly satisfactory reverberation time of 1.2 seconds, but at 128 hertz it could have a reverberation time of 7.2 seconds and at 2048 hertz, 0.6 second. Such a room would have

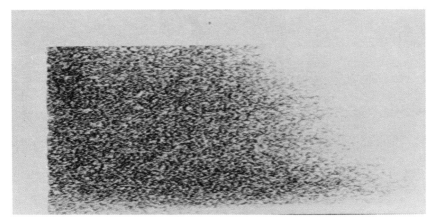

Figure IV-6. The report of Fig. IV-4 sounding in a highly reverberant corridor.

too little absorption for the bass notes of music, which would sound objectionably "boomy," too much absorption for the high frequencies of speech. In the record of Figure IV-5 the sounds in some frequency regions reverberate much longer than others; accordingly the reverberation time of that auditorium varies significantly with frequency. Sound spectrograms thus furnish a way of ascertaining the reverberation characteristics of a room at all frequencies of interest.

When the same gun is fired in a highly reverberant corridor that has not been acoustically treated, the sound reverberates for a very long time (Fig. IV-6). Again, as in the auditorium case, the reverberation time depends on the frequency of the sound.

We have already noted that the sound spectrograph was originally developed for analyzing speech by making the important characteristics of speech "visible." In speech analysis knowledge of the location of the significant broad areas of frequency is often more important than information gained from a narrow frequency analysis (e.g., one that shows the position and strength of the individual harmonics). For this reason two filter widths are always used in the sound spectrography of speech. One width encompasses rather broad frequency areas; it in effect suppresses information concerning the position and strength of the individual speech harmonics and delineates the strong broad frequency areas that are important in speech analysis. Because of its broadness, this filter also responds more rapidly than a narrow one, and the sudden starts, stops, and changes in speech are more accurately displayed when it is used.

Figure IV-7 shows how the use of the broad filter can be helpful in

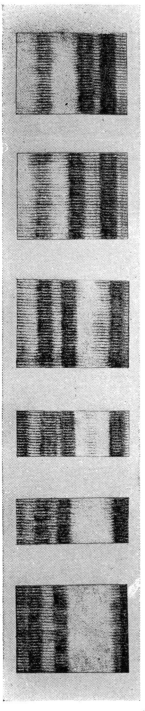

i (*eve*)   I (*it*)   e (*hate*)   ε (*met*)   æ (*at*)   a (*ask*)

Figure IV-7. Spectrograms of spoken vowel sounds. Preceding the words (which indicate the sounds) are phonetic symbols, which indicate a given sound, such as the *ee* in "eve" (designated by the symbol *i*).

speech analysis. The spectrograms portray six different vowel sounds. These are all voiced sounds, which means that they are generated by the vocal cords as tones comprising a fundamental frequency and numerous harmonics.

A narrow-band-filter analysis would portray the harmonic structure of vowel sounds by indicating the many individual harmonics as multiples of the fundamental frequency of the vocal-cord sound. The broad filter used in Fig. IV–7 cannot resolve the individual harmonics; it portrays only the loud areas in the frequency scale. The loudness of these frequency regions is caused by resonances in the human throat and vocal tract—an inherent property of their physical structures; harmonics falling within these resonant frequency areas are louder than others.

These resonances are called variously vocal resonances, formants, or bars. Speech analysts have ascertained that these formants, or bars, give strong clues to the voice sound being uttered; our way of understanding what sounds and words are being uttered by another person may thus, for voiced sounds, be intimately connected with the positions of the formants of these sounds in the frequency scale. Figure IV–7 shows how the formants of various voiced sounds occupy different positions in the (vertical) frequency scale.

The spectrogram at the far left of Fig. IV–7 portrays the sound *ee* as in "eve." Proceeding to the right, the vowel sound changes first to the *i* in "it," then to the *a* in "hate," etc., and alterations occur in the positions of all the formants.

The movement of the two lowest frequency formants in this series is rather interesting. The lower of these two first moves up in frequency as the higher one moves down, but in the right-hand spectrogram both move downward together. The position of these two formants in the frequency scale thus provides a rather significant clue to the identity of a particular vowel sound. Furthermore, because it is determined by a particular jaw and tongue position, the formant remains fixed in its frequency position even though the pitch of the vowel sound (determined by the vocal cords) may change. The broad-frequency-band filter thus suppresses the unwanted individual harmonic information and displays clearly the more significant formant location regardless of pitch.

V   Some Voice Patterns

The visible-speech patterns in Fig. IV-7 are patterns of steady vowel sounds—sounds that do not change throughout a record. One valuable feature of the visible-speech method of presenting the structure of speech is the portrayal of the time variations that are continually occurring in spoken words. Let us observe some of the more important *transient* forms of speech sounds.

Portrayal of Speech Sounds

Voiced and Unvoiced Fricatives.   The hissing-type speech sounds, such as the *s* in "sister" and the *sh* in "she" are called sibilants or fricatives (sibilant from *sibilare,* to hiss, and fricative from *fricare,* to rub, referring to the frictional sound of breath passing a narrowed point in the vocal tract). In the *sh* in "she" the vocal cords are silent, and this sound is therefore termed an unvoiced, or voiceless, fricative continuant. The same frictionlike sound can also be uttered with the vocal cords active; it then becomes the *z* in "azure" and is called a voiced fricative continuant. A similar change occurs for the *s* sound in "sister" (also a voiceless fricative continuant) when the vocal-cord sound is added to it, as, for example, in the word "zoo." This *z* sound is also called a voiced fricative continuant.

The characteristics of these two sounds are portrayed in the two visible-speech records of Fig. V-1. At the top is the analysis of the word "sue," comprising the unvoiced *s,* followed by the (voiced) vowel sound *oo.* The arrows indicate the most typical portion of each of the two sounds; between these arrow points there is a transition between the two sounds. The hissing sound of the voiceless *s* is noiselike (nonperiodic), and its spectrum is seen to extend over the entire frequency spectrum portrayed. When the sound changes to *oo,* the characteristic bars, or resonances, of a vowel sound become evident (which we saw in the last chapter) and the noiselike sound is replaced by a voiced one. For a short

42

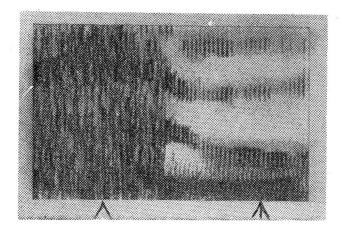

sue

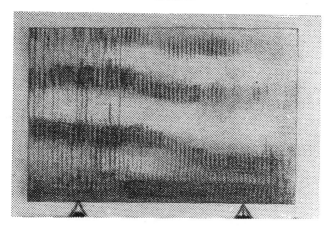

zoo

Figure V-1.   Spectrograms of the words "sue" and "zoo."

period the voiceless *s* sound is followed by a combination sound that possesses both resonance bars and the broad-frequency-band character-istic of the *s* sound to its left. The wide-band portion of this finally dis-appears when the sound changes to the purely voiced sound *oo* at the right.

The bottom portion of Fig. V-1 shows the vocal-cord (voiced) sound of "zoo." The voiced sound, with its accompanying resonances, is manifest

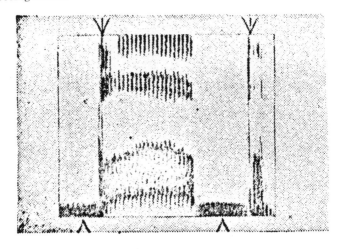

bob

Figure V-2. Spectrogram of the word "bob."

in both sounds (*z* and *oo*) as well as in the transition region between them. The *z* sound includes both the noiselike, broad-frequency-band characteristic of the voiceless *s* sound seen in the analysis of "sue" at the top and the resonances of a voiced sound. The wide-band characteristic dis-

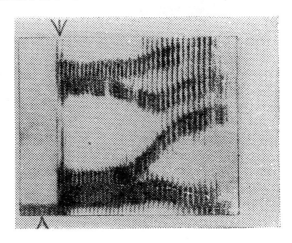

boy

Figure V-3. Spectrogram of the word "boy."

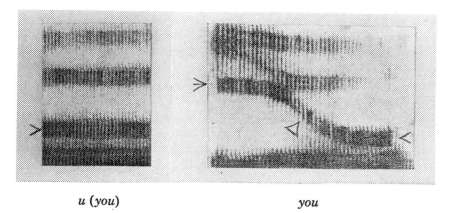

u (you)   you

Figure V-4. Spectrograms of the sound *oo* (indicated by the phonetic symbol *u*) of "you" and the complete word "you."

appears when the sound changes to the purely voiced sound *oo* at the right.

Stop Sounds. The word "continuant" in the expression "voiced or unvoiced fricative continuant" differentiates the continuous sounds of speech from those that require a cessation of sound for their formation. Such sounds are called stop sounds; they are caused by the complete interruption of breath by a stopping of the breath flow. This interruption can be generated at various places in the vocal tract, such as in the deep throat, in the middle throat, or at the mouth extremity, the lips.

An example of a stop sound is presented in Fig. V–2, which shows the analysis of the word "bob." Because the breath is completely stopped for the two *b* sounds, no high-frequency sound is emitted from the mouth (or nose) and there is therefore a blank space in the upper portion of the pattern. Similar stoppages occur for other sounds, such as the *ck* in "back," the *t* in "hit," and the *p* in "up."

Diphthongs. A speech sound that is a transition between two vowel sounds is called a diphthong. Figure V–3 shows an analysis of the word "boy," in which a diphthong is preceeded by the stop consonant *b*. The vowel sound changes rapidly from *oh* to *ee* to form the *oy* portion of the word. As in the word "bob" of Fig. V–2, during the stop sound of *b* there is a blank area that suddenly changes to a pattern with the characteristic resonances of a vowel. In this case, however, the resonances assume new positions (frequencies) in the smooth, fairly gradual way that is characteristic of a diphthong. The resonances are seen to glide from the *oh* position to the *ee* position during the transition period.

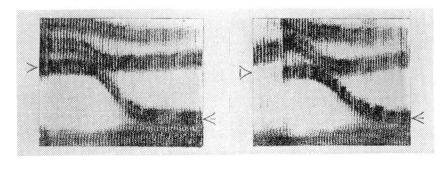

Iu (*new*)   *new*

Figure V-5. Spectrograms of the diphthong *ee-oo* (indicated by the phonetic symbol ɪu, as in "new," and the complete word "new."

The sound *ee* can also be combined with the sound *oo* to form diphthongs. This is shown in Fig. V–4. The *oo* sound (indicated by the phonetic symbol *u*) is portrayed at the left, and the change from *ee* to *oo*. as in the word "you," is shown at the right. Here the frequencies of the resonances change in a downward direction, in contrast to the rising pattern of "boy" in Fig. V–3.

Figure V–5 portrays the same diphthong by itself on the left and following the consonant *n,* thus forming the word "new," on the right. The consonant *n* is voiced, as shown by its bar, or resonance, pattern. It suddenly changes to the purely voiced sound *ee,* which in turn undergoes a more gradual transition to the sound *oo,* forming the diphthong.

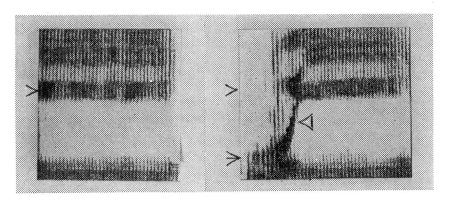

i (we)   *we*

Figure V-6. Spectrograms of the *ee* sound (indicated by the phonetic symbol *i*) of "we" and the complete word "we."

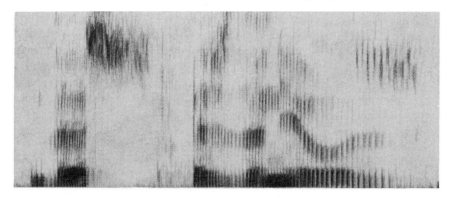

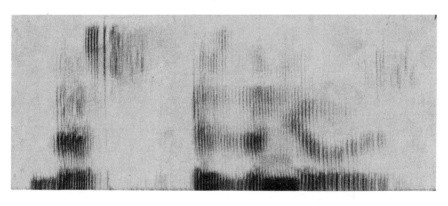

THIS   IS   THE   NEWS

Figure V-7.   Spectrograms of the sentence "This is the news": top, the original speech; bottom, a vocoder re-creation.

When the vowels in the diphthong *ee-oo* are reversed, another diphthong is formed, the one in the word "we." Figure V-6 shows on the left the final sound *ee* and on the right the diphthong *oo-ee,* which is the word "we." The gradual glide in the resonance positions is again evident.

Portrayal of Synthetic Speech Sounds

Much effort has been devoted in recent years to devising ways of generating synthetic speech. It is known that for speech to be understood and for the talker's voice to be reasonably well recognized the transmission system (e.g., a telephone circuit) must have a frequency response that enables it to transmit without distortion signals whose frequencies range between 300 and about 3500 hertz. (For true fidelity this band should extend from 100 to 12,000 or 15,000 hertz.) It is also known, however, that even the telephone bandwidth is much wider than that necessary to transmit the information content of the speech signal. Accordingly, if speech is processed by coding at the sending end and reconstructed at the receiving end, rather good results can be obtained with a transmission channel of much narrower bandwidth than that usually used.

The Vocoder.   One of the more successful of the voice-coding methods is called the vocoder, a device largely attributable to Homer Dudley and his co-workers at the Bell Telephone Laboratories. Because they clearly depict the most important characteristics of speech, the voice-portrayal techniques discussed in this chapter are often used to display how well the vocoder or any other speech synthesizer can reconstruct original speech sounds.

Figure V–7 is a display of the sentence "This is the news." The top of the figure shows the analysis of the original speech; the bottom, a vocoder re-creation of the sentence. As the close similarity of the visible-speech spectrograms suggests, the two versions sound very much alike. The coded speech, however, has the potential of being sent over a transmission circuit of much narrower bandwidth. The vocoder employed in this record possesses certain improvements first suggested by R. L. Miller of the Bell Telephone Laboratories.

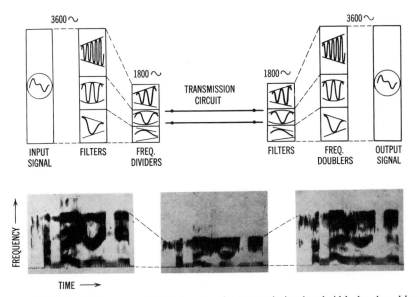

Figure V-8.   The Vobanc, a method of conserving transmission bandwidth developed by Bruce P. Bogert of the Bell Telephone Laboratories.

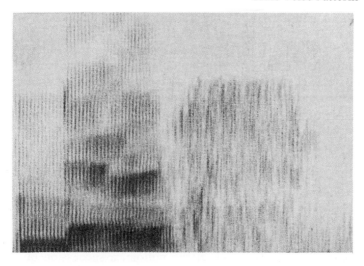

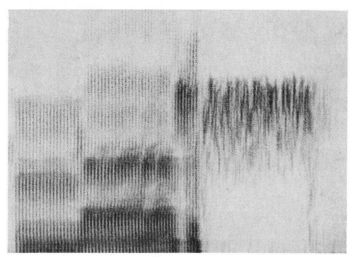

Figure V-9. Spectrograms of the word "nurse": top, the spoken word; bottom, a succession of four speech sounds simulating the word.

The Vobanc. Another procedure for transmitting speech over a transmission circuit whose bandwidth is narrower than that usually employed is a technique called the Vobanc. It was investigated by Bruce P. Bogert of the Bell Telephone Laboratories. Figure V–8 indicates the method employed in this technique.

The original speech sounds (at the left of Fig. V–8) are passed through three filters, and the output frequencies of each are divided in half. The original 3600-hertz band is thus reduced to 1800 hertz. The narrower band is thus reduced to 1800 hertz. The narrower band speech sounds are then transmitted over a narrower band circuit (the analysis of these is shown at the bottom center of the figure), and at the receiving end three filters are again used, their output signals being doubled in frequency. Because the visible-speech analysis at the lower right shows moderate similarity to the analysis of the original speech at the left, one can conclude that the distortion introduced by this procedure is not extensive.

The Phoneme Vocoder.   Synthetic speech can also be generated without utilizing an original sample. Switching keys are made to correspond to particular speech sounds and are activated by an operator; alternatively a computer is programmed to cause the prearranged sounds to be heard in the proper succession. Figure V–9 shows, at the top, the word "nurse" as spoken and, at the bottom, a succession of the four speech sounds (phonemes) *n, uh, rr,* and *sss.* Again, a modest resemblance between the two is seen.

This process can also be used as a vocoder; at the sending end each phoneme is electronically recognized in succession and immediately (and synthetically) sounded at the receiving end. Very large savings in bandwidth can be attained by this procedure, but all of the characteristic quality of the original speaker is, of course, lost.

Problems in Reading Visible Speech.   A person could conceivably learn to read the visible-speech spectrograms of spoken words as one reads a printed page. This could permit a deaf person to understand speech if it were continually presented to him visually and if he could learn to read the presentation.

Figure V–10 indicates one of the problems in this regard. The three spectrograms are all of the same vowel sounds, the *a* in the word "at." On the left the analysis shows the sound as spoken by a man, at the center as spoken by a woman, and at the right as spoken by a child.

The three spectrograms show different frequencies of the resonances since these are determined in part by the sizes of cavities in the vocal tract, which vary from person to person. This variation in the resonance frequencies is one aspect that makes it difficult for a viewer to interpret visible speech infallibly.

Dynamic Spectrograms.   In an attempt to make visible-speech patterns more recognizable the usual speech-analysis procedure was modified by accentuating the regions in which changes occur.

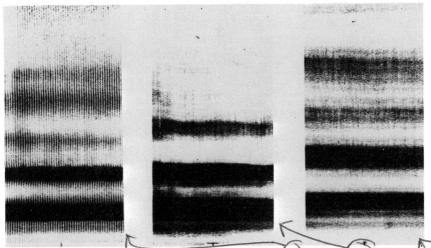

Figure V-10.   The sound of *a* in the word "at," as spoken by, left, a man; center, a woman; and right, a child.

Figure V-11 shows, at the top, three utterances of the word "bird," analyzed in the manner we have discussed (normal, wide-band, spectrograms). In the lower series the same utterances are presented as *dynamic* spectrograms. In each of these, at the left, there are black areas that indicate the change from the stop sound *b* to the two resonances (bars) of

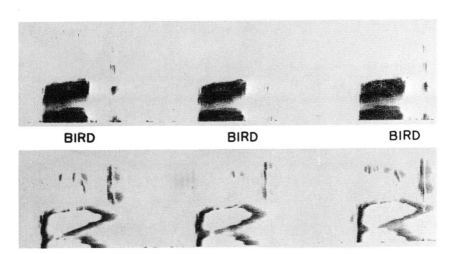

Figure V-11.   Dynamic spectrograms of speech.

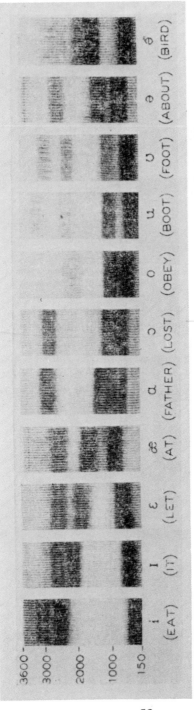

Figure V-12. Sounds generated synthetically by the electrical vocal tract, developed by H. K. Dunn of the Bell Telephone Laboratories.

the vowel sound, but immediately thereafter, because the resonances remain fairly constant (as seen in the upper series), no further heavy marking of the two resonances occurs. The record does show, however, that the upper resonance is rising and the lower one is (later) falling in frequency. It also shows clearly that the *d* is a voiced sound (where the upper records are completely blank), and the ending *duh* of the *d* is more easily recognizable.

Electrical Vocal Tract. Figure V–12 shows two sets of vowel analyses, the lower set being of spoken vowels and the upper being of sounds generated by a device developed by H. K. Dunn of the Bell Telephone Laboratories and called an electrical vocal tract. This device substitutes an electrical transmission line for the acoustic transmission line or vocal tract of the throat and permits electrically varying certain parameters, such as the opening of the lips and the position of the tongue hump. As the analyses show, this device produces a very close duplication of the vowel sounds.

Computer Speech. Recently digital-computer technology has begun to play an important role in speech analysis and synthesis. The spectrum, or Fourier, analysis of a speech wave, which has utilized the filtering techniques developed by Ralph Potter, can now be performed by computer. This is accomplished by taking samples of the wave form, as indicated for one complete period of a wave in Fig. V–13. From this information a very close approximation to the spectrum analysis of the wave can be arrived at. (Because only discrete points are sampled, it is

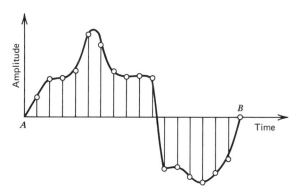

Figure V-13. A sampled wave form.

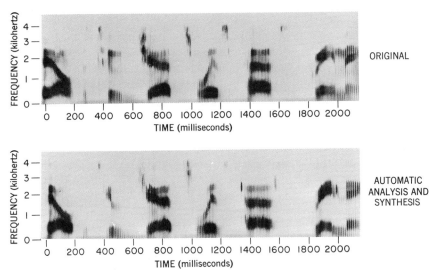

Figure V-14. Spectrograms of the sentence "High-altitude jets whiz past screaming." Top, original human speech; bottom, computer-synthesized speech.

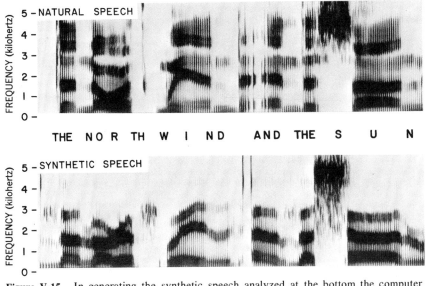

Figure V-15. In generating the synthetic speech analyzed at the bottom the computer actually "read" a printed text and, with the help of a few instructions, was able to match natural speech very closely.

54

called a *discrete* Fourier analysis or, for waves whose spectrum is not constant with time, a discrete Fourier *transform*.)

Obviously such a sampling procedure fits in nicely with step-wise, or *digital,* computers, and soon a way of programming for Fourier transforms was developed. The procedure now employed most widely, because it avoids a large number of time-consuming operations, is called the "fast Fourier transform" or FFT for short. The FFT procedure also permits the pitch of a voiced signal to be acquired, and with these two information items, the pitch and the spectrum, ways have been found to make computers generate sounds that match human speech very closely.

One of the outstanding contributors to this field, J. L. Flanagan of Bell Laboratories, describes the process as follows:

Individual words spoken by a human are analyzed, converted into numerical information, and stored in a computer. Preprogrammed instructions tell the machine to link the stored data into the numerical equivalent of sentences and then convert this digital information into synthetic speech.

Figures V–14 and V–15 show spectrograms of the original human speech and of the computer-reconstructed synthetic speech.

VI   Some Music Patterns

The technique of portraying the structure of sound as developed by Potter and his associates for speech sounds is also useful in the analysis of many musical sounds; a few examples of these are presented and discussed in this chapter.

Vibrato and Tremolo

A periodic variation of approximately 5 hertz is usually imparted to an instrumental or vocal solo voice. This variation is called a *vibrato* if the frequency is varied and a *tremolo* if the amplitude is varied.

The frequency of the tones generated by a pipe organ is usually relatively fixed because the frequency of the tone is generally determined by the pipe length. For a pipe-organ note, therefore, only an amplitude change is usually possible—that is, a tremolo. The tones of string instruments, such as violins and cellos, on the other hand, can be given a frequency change (vibrato) through the periodic shortening and lengthening of the string being bowed, as accomplished by the performer's periodic finger-and-wrist motion.

Either an amplitude tremolo or a frequency vibrato, or both, can be imparted to a note sung by the human voice. The more pleasing of these is

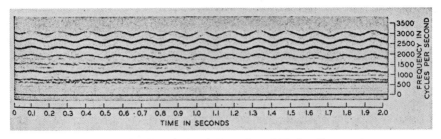

Figure VI-1. A narrow-band analysis of the voice of Enrico Caruso shows a strong periodic frequency variation corresponding to a vibrato.

56

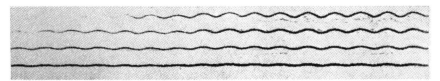

Figure VI-2. A record of the soprano Lily Pons. The wide separation of the harmonics indicates that the note being sung is very high in pitch.

generally felt to be a pure frequency variation, and singers therefore try to achieve such a vibrato. Figure VI–1 is a narrow-band analysis of the voice of tenor Enrico Caruso, a singer famous for his powerful operatic voice. The periodic frequency variation (vibrato) is evident. A similar record of a vocal selection sung by the soprano Lily Pons is shown in Fig. VI–2. A frequency variation is again evident, and there is practically no accompanying amplitude variation. It is noted that during the course of this record the intensity of the third and fourth harmonics increases.

Musical Instrument Formants

The numerous vowel-sound analyses presented in the preceding chapter show the characteristic vocal resonances, or formants, which are made more prominent by the use of broad-band filters in the analyzing process. Since they change as the characteristic sound is altered from one vowel to another, the formant positions in the frequency scale are generally associated with the characteristics of the sound.

Many musical instruments have resonance cavities or resonant structures, and, because their originally generated sound is usually rich in harmonics, these resonant structures accentuate or amplify the harmonics that fall within the frequency band of the resonance. It is generally agreed that the characteristic sound distinguishing one musical instrument from another is heavily influenced by the presence of these resonances. Experiments have placed the formant for the trombone at 1020 hertz, the important one for the French horn at 1350 hertz, and one for the trumpet at 1520 hertz. Here we define the formants in musical instruments in the same way as the voice formants — namely, as strong resonance regions that reinforce the harmonics falling within them.

Figure VI–3 shows three spectral analyses of a bassoon tone played at three different pitches. At the higher pitches the harmonics are spaced further apart, and in each case the harmonics of all pitches fall into a strong resonance located at about 500 hertz (0.5 kilohertz).

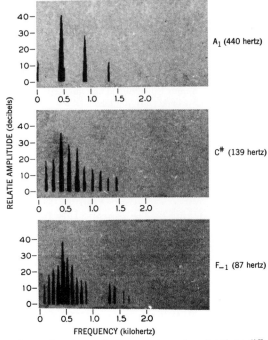

Figure VI-3.   Spectral analyses of a bassoon tone played at three different pitches.

The magnitude of the resonance effect in this case is quite comparable to that of vocal resonances. On passing through the resonance point the harmonics undergo 20- to 30-decibel variations in strength. This effect is also exploited in the process of imitating certain organ and orchestral sounds in electronic organs. The method employs an electrically resonant circuit whose frequency position and bandwidth correspond to those of the instrument to be imitated. When any of the organ's oscillator tones, which are quite rich in harmonics, are passed through this circuit, the harmonics that fall within the resonance region are reinforced and the desired instrument tone is generated, irrespective of the pitch of the original oscillator tone. As seen in Fig. VI-4, several of the tone-color circuits of an electronic organ, with which the author was involved (the Baldwin electronic organ), include inductance–capacitance combinations; these provide the desired band-pass circuit.

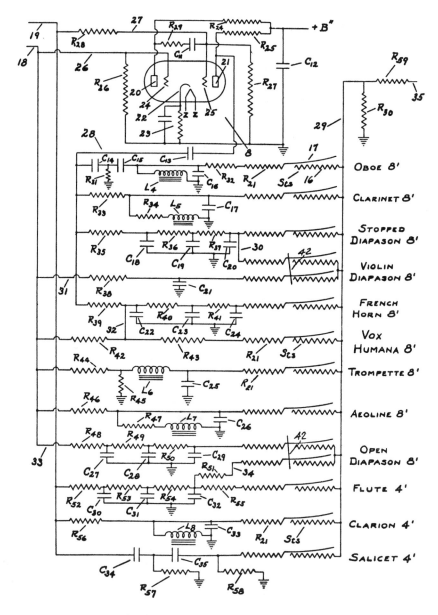

Figure VI-4. Tone-color circuits employed in an electronic organ often include resonant circuits to produce the effect of formants. In this figure the oboe, clarinet, trompette, aeoline, and clarion stop circuits include inductances. (From a Baldwin organ patent issued to the author).

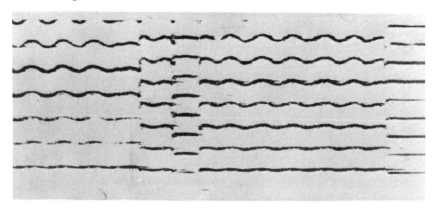

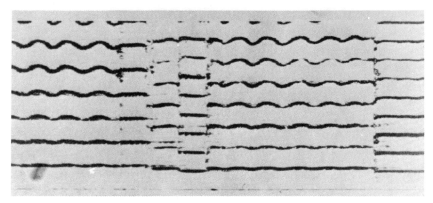

Figure VI-5. Sound spectrograms of a selection played by Zino Francescatti on a Guersan violin (top) and on an inexpensive violin (bottom). The frequency scale is 0 to 6500 hertz.

Instrument Quality and Visual Presentation

One of the problems that has confronted musical instrument analysts is ascertaining the reasons for the recognized quality of an instrument. This has been particularly troublesome in the case of violins. For many years violinists have continually favored violins of certain old violin makers, such as Stradivarius, Guarnerius, and Guersan, yet the reasons for their acceptance have not been discernible in scientific comparison tests with ordinary violins.

The visible-speech procedure may help analysts in ascertaining the rea-

Figure VI-6. Zino Francescatti playing the violin selection analyzed in Fig. VI-5, with the author standing behind him.

sons for the popularity of a particular instrument and its presumed tonal superiority. Figure VI–5 presents the narrow-band analysis of a sequence played by the violinist Zino Francescatti in an anechoic room at the Bell Telephone Laboratories at Murray Hill, New Jersey (see Fig. VI–6). The bottom analysis portrays the sequence played on an inexpensive violin, and the top one portrays the same sequence played on a

Guersan violin. These records show that the inexpensive violin has stronger low-frequency harmonics both at the left of the spectrogram and at the far right. Furthermore, the inexpensive violin generates more extraneous noise between the harmonics during passage from one note to the next. One might expect that on rapid passages this extraneous noise would diminish the clarity of tone.

Formant Duplication by Organ Mixtures

Because formants have much to do with the characteristic timbre of musical instruments, a comparable effect is often sought in the design of organ stops referred to as "mixtures." In such stops numerous pipes are made to sound even when only a single key is depressed. For the more usual stop, such as, for example, a diapason, only one pipe speaks when one key is depressed, and hence, only one set, or rank, of pipes (one for each key of the keyboard) is required. For a mixture a plurality of ranks is needed, and the term applied to the mixture is dependent on the number of pipes that speak (e.g., a four-rank mixture).

Mixtures that are designed to simulate formants accomplish the effect by the expedient of breaks; that is, the lower octaves have the very high partials of the depressed key represented, whereas for the higher octaves lower partials (harmonics) are made to sound. Thus, as one goes up the scale, the breaks remove upper harmonics and add lower ones. These breaks thus impart brilliance to the lower octaves, richness to the middle octaves, and fullness to the upper octaves. Helmholtz compared the resonances, or formants, of a violin to the effect of mixtures (compound stops) in the organ:

The partial tones of the strings [of a violin] are reinforced in proportion to their proximity to the tones of the resonance box. The deepest notes of a violin will have their octaves and fifths favored by resonance, whereas the higher notes will have their prime tones assisted. A similar effect is attained in the compound stops of the organ by making the series of upper partial tones, which are represented by distinct pipes, less extensive for the higher than for the lower notes of the stop.

To illustrate how this effect is achieved let us examine a few mixture specifications. Figure VI–7 shows the arrangement of a three-rank cymbal mixture with six breaks; it is a stop in the Aeolian Skinner (G. Donald Harrison) organ at Christ Church, Houston. In this figure the long continuous line represents the key that is depressed, and that note corresponds to the fundamental. Thus at the very lowest key of the keyboard, corresponding to low C (i.e., CC) this three-rank mixture sounds three notes (three harmonics) encompassing C^3 to C^4. At the very highest key of the keyboard, C^4, the mixture sounds C^5, the first overtone (the second harmonic) of C^4, as well as the third and fourth harmonics. The harmonics generated lie in approximately the same region of pitch no matter where the fundamental lies, that is, between C^3 and C^6. They thus correspond to the action of a fixed resonator producing a formant. It may be noted in the chart that the harmonics shift slightly higher in pitch with a higher fundamental; this effect is also observed in the formants of orchestral instruments.

Visible Presentation of the Chorus Effect

A relatively recent addition to the family of musical instruments is the electronic organ. In this instrument transistors generate and amplify the musical tones, which then emerge from a loudspeaker. Significant economies are achievable with this organ because the tone generated by a single electronic oscillator can be given many different tonal qualities, including diapason, string, reed, and woodwind tones. The pipe organ, on the other hand, requires a complete rank of pipes (one for each note of the organ keyboard) for each tone color desired. Hence organs that pro-

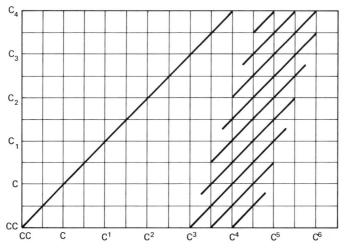

Figure VI-7. Design of the three-rank cymbal III mixture of the organ at Christ Church in Houston.

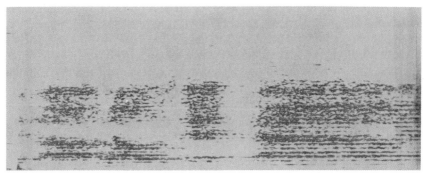

Figure VI-8. Spectrogram of a chorus singing the words "Thou knowest it telling" from the Christmas carol "Good King Wenceslas."

vide many different tone colors as registration possibilities have a very large number of pipes.

We noted in Fig. VI–4 that the tone-color process of one form of electronic organ utilizes an oscillator tone that has a strong harmonic content. Numerous different instrumental tones can then be produced by the use of many different electrical filters. Although an economy in oscillator numbers is thereby achieved, one disadvantage results: the organ is incapable of providing a chorus quality or chorus effect. It is this chorus effect that causes the sound of a vocal quartet to be distinctly different from that of a chorus of voices singing the same four-part harmony.

This characteristic sound of a chorus is due to the frequency-broadening effect caused by the slight deviations in pitch generated by many individual voices. This difference also exists between a string quartet and a string symphony playing the same selection. Now a pipe organ also generates a chorus effect when many stops are drawn because the many pipes sounding in unison are similarly slightly different in pitch. This characteristic is difficult to duplicate in an electronic organ, in which many tone colors are provided by a single electronic oscillator.

The characteristics of the chorus quality can be displayed by the visible-speech-portrayal technique, and this can help designers in seeking ways of duplicating this effect in electronic instruments. Figure VI–8 is a record of a chorus singing the words "Thou knowest it telling" from the Christmas carol "Good King Weneceslas," made by the narrow-band analysis procedure used for the solo-voice records of Figs. VI–1 and VI–2. The individual harmonics in the vowel sounds sung by the chorus are not at all as sharp or as clear as those in the solo-voice records. The existence of many voices, all very slightly different in pitch, has produced a spreading out of the harmonics.

Because the broadening just noted is reminiscent of the broadened-frequency pattern of a noiselike sound, it suggests that the use of noise-like electronic generators instead of pure-tone oscillators could be advantageous in electronic organs, as shown in Fig. VI–9. In the diagram at the top simple electronic filters, one for each key, replace the individual elec-

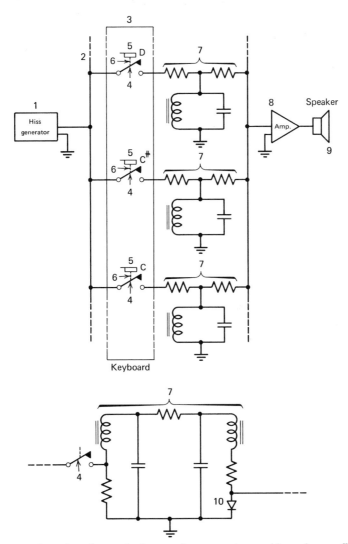

Figure VI-9. Procedure for employing a noise source to provide a chorus effect electronically. (From a patent issued to the author).

tronic oscillators of the organ. They are narrow enough in bandwidth to provide a tonal quality comparable to that of the spread-out frequency bands shown in the chorus harmonics of Fig. VI–8. This simple version, however, generates only pure, fundamental tones because the filters pass only a narrow band of noise surrounding the desired note; the resulting tone is free of harmonics. To achieve a chorus effect with tones that are rich in harmonics the pure, filtered tones are rectified, as shown in the bottom diagram of Fig. VI–9. A large harmonic content is imparted to the tone, and these harmonics resemble those of Fig. VI–8 in their noiselike quality.

VII   Other Sound Patterns

Correlation Patterns

Another way of portraying the structure of sound is through the use of a process called correlation. Consider two identical replicas of a wave having a certain amplitude variation with time. If the amplitudes of these two replicas are multiplied together, a positive product will always be obtained: when both are in a positive excursion, the product of two positive values is positive, and when both are in a negative excursion, the product of the two negative values is again positive. Accordingly, if the resulting positive product is continuously added, by having, for example, both wave sections passed successively through a multiplier and then an adder (an integrator), a positive value continuously increasing with the length of the wave sample is obtained.

In the actual, real-time case the wave amplitudes vary very rapidly with time; however, the multiplication of the two wave amplitudes can also be accomplished rapidly by electronic procedures, as can the continuous summation (integration) of the resulting positive products. Accordingly the continued electronic multiplication of two highly similar waves and the continued summation of the multiplied signals eventually results in a very large summed result. This process of multiplying and adding two identical waves is referred to as the correlation process.

We consider now the correlation process that results when the two waves are identical but are multiplied and summed with the time coordinate of one of the waves *varied* relative to the other. We consider first two identical single-frequency (sine) waves. When they are exactly aligned (i.e., when the time coordinates are identical), the multiplication and summation continues to yield a positive result, as in the case of the non-uniform waves discussed above. However, when the time coordinate of one wave is displaced in relation to the other, full summation results only when the waves are displaced by one full wavelength or by an integral number of wavelengths. If now the time displacement is made *continuously* (and uniformly), the correlation (the result of the multiplication and

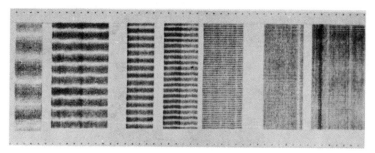

Figure VII-1.   Correlatogram of a sine wave.

summation) of the two waves will alternate between a positive *maximum* (when the products of the two pluses and the two minuses all add) and a negative *minimum* (when the pluses are multiplied with the minuses and the minuses are multiplied with the pluses).

The visual portrayal of the result of this process for a sine wave is shown in Fig. VII–1. The relative variation of the time coordinate of the two waves is plotted vertically, and the real-time variation of the wave itself is plotted horizontally. Because in this case the example is a continuous (single-frequency) sine wave, there is no variation in the horizontal direction; that is, the positions of the horizontal bars do not vary with time. Following the practice of calling visible-speech frequency analyses spectrograms, this portrayal carries the name "correlatogram."

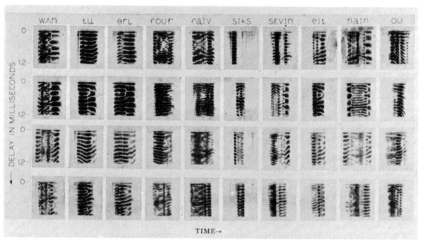

Figure VII-2.   Correlatograms of the 10 digits (indicated by phonetic symbols at the top) as spoken by four different persons.

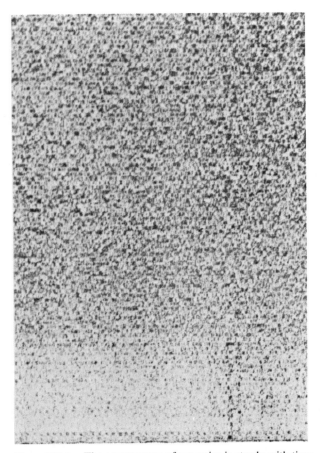

Figure VII-3. The spectrogram of sea noise is steady with time.

When a wave whose amplitude varies in a random fashion is subjected to the time-varying autocorrelation process, the correlation "bars" are not constant with time, nor are they in general as uniform in the vertical display pattern as those in Fig. VII-1. Speech sounds are waves of this type since, as we have seen in the spectrograms of speech, they rapidly change from the noise or hiss type of sounds to voiced, or vowel, sounds, and even the latter change rapidly in the frequency of the fundamental tone. Figure VII-2 shows correlatograms of the 10 digits as spoken by four different persons.

Sounds in the Sea

Sound waves also propagate in water, and a technology called sonar (by analogy with the electromagnetic wave technology called radar) exploits underwater sound propagation to determine the presence and location of objects on or beneath the surface of the sea. In analyzing these and other sounds in the oceans a visual portrayal is often of advantage. We shall here examine the use of portrayal techniques for underwater sounds.

Sea Noise. Wind and the storms caused by high winds generate a a noiselike sound in the oceans. The frequency spectrum of this sea noise is very broad, comparable to the hissing noise of wind through trees or to the hiss sound of speech sibilants. A visual plot of this noise is shown in Fig. VII–3. It is noted that the record is steady with time and has a smooth frequency structure.

When the noise of a storm is intense and moderately well localized, as in a typhoon or hurricane, directional underwater listening devices can track the movement of the disturbance. Figure VII–4 shows records

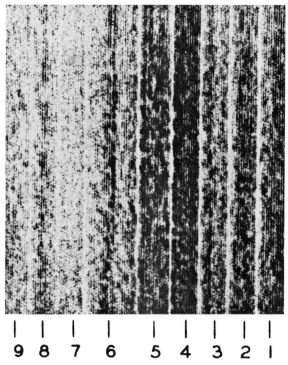

Figure VII-4. A set of nine spectrograms obtained by pointing a directional underwater acoustic receiver in nine different directions. A hurricane is located in the directions showing the blackest records.

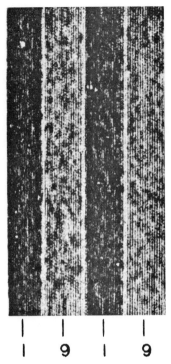

Figure VII-5. Alternately pointing the receiver of Fig. VII-4 toward the hurricane and toward a quiet area shows the difference in noise levels.

made by an underwater listening device aimed in nine different directions, the intensity of the sea noise being greatest in the directions at the right.

Figure VII–5 similarly portrays alternate signals received from only two directions. One is the direction in which the eye of the storm lies; the other is an area of the ocean not yet affected by the storm. The second and fourth records show a far weaker signal, indicating that the sea noise in that direction is small. The first and third records are much darker, showing that the noise level in that direction has been greatly augmented by the storm.

Propeller Noise. A propeller-driven ship also generates a broadband form of noise by virtue of the phenomenon called cavitation. A rotating propeller creates both positive and negative pressures in water, and the extremely low negative pressures cause the formation of tiny bubbles that collapse almost instantly thereafter. Each collapse resembles a tiny explosion, and the noise thus generated, like the noise of the shot discussed in Chapter IV, is of the wide-band type. The process of the myriad bubbles collapsing is the cavitation, resulting in the continuous generation of wide-band noise. Since it comes from the propeller, this noise, like

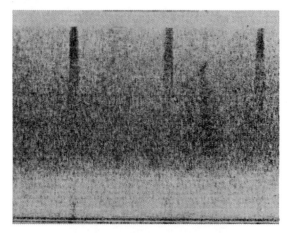

Figure VII-6.   Noise made by a sea creature sometimes referred to as "Barney's Beast."

the eye of a storm, is localized, and its direction can be similarly detected and localized by listening devices.

In listening to the cavitation sound of a ship's propeller one perceives a rhythmic pattern caused by the cavitation's being greatest near the water surface. The operator of a listening device can actually, by counting the number of rhythmic beats per minute, estimate the speed of the ship (provided he knows the propeller and ship characteristics). A second way of determining the rhythmic count is to introduce the visible-speech analyzer into the listening system.

Animal Sounds.   Many fish and other sea animals make noises that can be picked up by underwater microphones (hydrophones). Whales produce a single-frequency sighing sound, and certain forms of shrimp create a snapping noise.

The fish or animal responsible for several sounds that have been detected has not yet been identified. One, known humorously as "Barney's Beast" (after Harold Barney, a Bell Laboratories scientist, who first proved that the sound source moved about), generates the sound shown in Fig. VII–6. Because of its rhythmic pattern, this sound, often lasting for several minutes, was at first believed to be a man-made, mechanical one (possibly from a pump or well-drilling rig). By using a correlation-pattern record, Barney was able to ascertain that the sound source moved and therefore was produced by an underwater animal of some sort. Years after it was first heard (and seen) the sound remains unidentified.

Another still unidentified underwater sound, produced by a moving source and also believed to be associated with an underwater fish or ani-

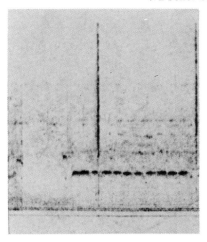

Figure VII-7. An underwater noise observed over extended areas of the Atlantic and Pacific oceans is strongly peaked in the vicinity of 20 hertz. It has been investigated and discussed by R. A. Walker of the Bell Telephone Laboratories.

mal, is shown in Fig. VII–7. This sound, which has a frequency of about 20 hertz, was analyzed by a Bell Laboratories scientist, R. H. Nichols, and has been detected in many areas of the Atlantic and Pacific oceans. It is a pulsed (on-and-off) sound that resembles a telegraphic code message on being recorded and played back at higher speed. Because of its low frequency, it does not attenuate very much with distance as the higher audiofrequency sounds do and hence can be detected at very sizable ranges.

Acoustic Holography

The most recently developed method for seeing sound is by means of the technique called holography. Originally invented as an optical process by the British scientist Dennis Gabor, in 1947, holography has more recently been extended to other fields, including acoustics and microwaves. In the latter two cases the objects of the "scene" are "illuminated" with sound waves or microwaves, and the reflected (or diffracted) wave field is converted to an optical or photographic record. This record is then illuminated, as in the case of a true optical hologram, with coherent (laser) light, and the original "acoustically illuminated" scene is reconstructed and made visible to the eye.

Principles of Holography. Light-wave holograms are most easily

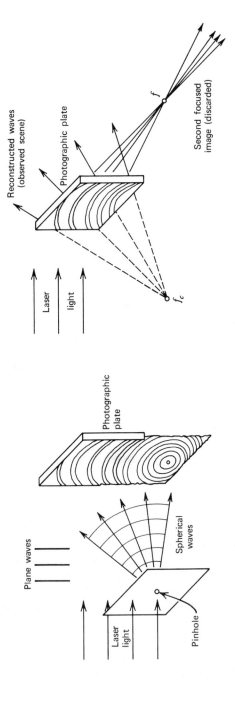

Figure VII-8. Circular interference fringes, corresponding to a zone plate, are created by the interference pattern formed by plane waves arriving from the left and by spherical waves emerging from the pinhole. If a portion of this pattern is recorded on a photographic plate, the resulting hologram, when illuminated, causes the generation of waves that reconstruct a virtual image of the pinhole at its original location.

explained as diffraction devices. Figure VII–8a, shows the photographic recording of the combination of two sets of light waves, one set being the plane (coherent) waves passing over the card with the pinhole (usually referred to as the reference waves), and the other the spherical waves passing through and issuing from the pinhole (the waves of interest in this case). Such a combination of spherical and plane waves generates, at the plane of the photographic plate, bright rings where constructive interference occurs. The spacing between these rings diminishes as the distance from the central ring increases; an actual photographic recording of such an interference pattern is shown in Fig. VII–9.

This pattern is identical with that of an optical diffraction device called a zone plate. The transparent rings transmit light energy that will add constructively at a "focal point," and the black, opaque rings block out energy that would cancel at that focal point. A cross section of a typical zone plate is depicted in Fig. VII–10, which also shows how to determine the position of the blocking rings.

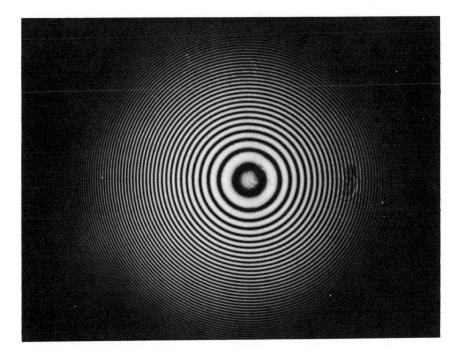

Figure VII-9. A photographically recorded interference pattern, generated by recombining coherent plane waves and spherical waves, resembles a zone-plate pattern.

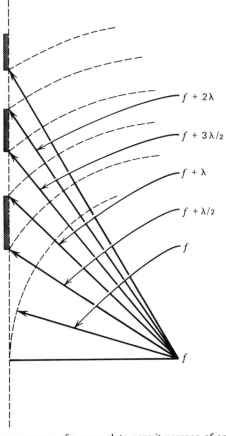

Figure VII-10.   The open spaces of a zone plate permit passage of energy that will add at the focal point f, and the opaque rings prevent passage of energy that would interfere at that point (λ = wavelength).

An analysis of the action of this structure when illuminated from the left by plane waves shows that it generates not only a set of focused waves (converging at the focus f) but also a set of diverging waves, appearing to emanate from a conjugate focal point f_c. Returning now to Fig. VII-8b, we see this double-action effect. Laser light illuminating the hologram (the photographic recording made at the left) causes a real image of the pinhole light source to be formed at the right (by virtue of the converging zone-plate waves) and also, at the left, a *virtual* image of the pinhole light. An observer thus imagines that a point light source is located at a fixed position in space behind the hologram. Had there been two pinholes

Figure VII-11. Three photographs of a hologram being illuminated with laser light. The hologram recorded a scene comprising three vertical bars; for these photographs (from left to right) the camera was moved successively farther to the right, finally causing the rear bars to be hidden by the front bars.

in the original scene, two zone plates would have been recorded on the photographic plate, and if this hologram were later to be illuminated with laser light, an observer would imagine that he saw two light sources.

Because any scene can be considered to be made up of many, many light sources, all of different brightness, a hologram photographically records the *superposition* of these many zone plates. In looking at the developed and reilluminated hologram an observer imagines that he sees,

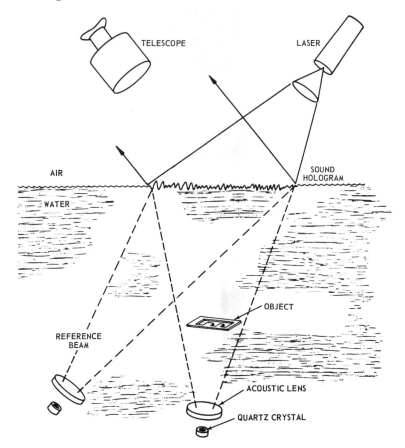

Figure VII-12.   Essential elements of a holographic sound-light image converter (after R. K. Mueller of the Bendix Research Laboratories).

standing out in space behind the hologram, the many light sources making up the original scene. The three-dimensional realism of optical holograms can best be demonstrated by showing several photographs of the image presented to a viewer by one hologram (Fig. VII–11).

Acoustic Holograms.   Since a hologram is a photographic record of the interference pattern generated between a set of waves of interest and a set of reference waves, it is obviously possible to make holograms of other forms of wave motion, provided the wave interference pattern can somehow be recorded. In Chapter II we saw that the wave-progression picture for sound waves could indeed be recorded photographically and that the process for doing this involved using a second set of waves as a

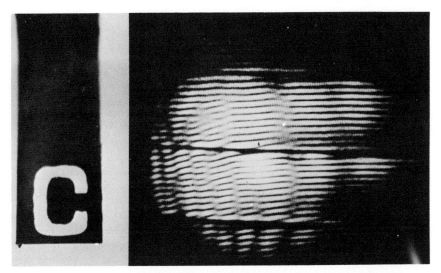

Figure VII-13. At the right is an acoustic hologram of the letter C, the original object being at the left.

reference set. Hence such images as those shown in Figs. III–14 and III–3 can be regarded as acoustic holograms, since they are recordings of interference patterns between waves of interest and a set of plane reference waves. The striations in these figures are acoustic fringes, exactly like the optical fringes that are formed when coherent light waves interfere.

For these fringe patterns no reconstruction process was performed, as the interest there was in the fringe pattern itself (the wave-progression pattern). Nevertheless, to record acoustic fringe patterns for hologram use the same technique is applicable, and acoustic holograms and their reconstructions have been made in this way in numerous laboratories.

Liquid-Surface Holograms. A real-time procedure for viewing acoustic holograms that bypasses the photographic recording process is shown in Fig. VII–12. Here the waves of interest are those diffracted by the object (the structure shaped like the letter E) originating at the bottom of the sketch, and their interference pattern, with the reference waves, at the liquid surface causes the surface to have ridges on it. Any liquid-gas surface is a pressure-release surface, and the stationary interference pattern of the acoustic waves mechanically deforms the surface in Fig. VII–12. The ridges correspond to the acoustic interference fringes referred to above, and reconstruction of the acoustic hologram image can therefore be accomplished by causing laser light to be diffracted by these ridges when it is reflected off the liquid surface. This real-time acoustic holog-

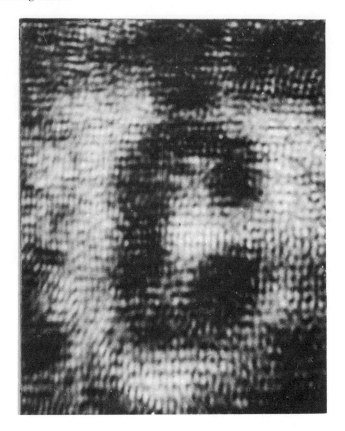

Figure VII-14.   Reconstruction of the hologram of Fig. VII-13. The outline of the letter C can be seen.

raphy procedure was first developed by R. K. Mueller of the Bendix Research Laboratories.

In underwater exploration the visibility near the bottom is often quite poor, so that optical methods are not feasible. Also, other acoustic techniques, such as those offered by sonar, may not be as useful as the results obtainable with acoustic holography. Figure VII–13 is a photograph of an acoustic interference pattern (the hologram) and Fig. VII–14 is the image that can be reconstructed by illumination with coherent (laser) light.

Superdirectivity

A visual presentation of certain sound-wave patterns often can make the phenomena responsible for the pattern more understandable. One such

Figure VII-15. Five loudspeakers spaced one-quarter wavelength apart can form a super-directive radiator if alternate speakers are reversed in polarity.

case involves a wave radiating or receiving process called superdirectivity. The theory on which this process is based predicts that a long-standing hypothesis in the field can be challenged. Proponents of superdirectivity admit that the availability of practical uses for it are quite limited, but even the mere existence of the effect, under any conditions, was for a time not acceptable to many scientists.

Classical optics argues that the sharpness of the beam of a radiator or receiver is determined by the ratio of the wavelength to the aperture size. For example, a line aperture or line array of length L has a beam width (to the half-power point) given by the expression 51 λ / 1 (where λ = wavelength). The superdirectivity theory states that an array of the same length can be endowed with a sharper beam than that defined by the above expression if certain polarity reversals are incorporated into the elements of an array.

Figure VII–15 shows a five-element array of loudspeakers. The array can be energized in parallel, all units having the same polarity, or polarity reversals can be imparted to any of the five still-paralleled units. Figure VII–16 shows the measured beam when all units are given the same polarity (the curve of the broader of the two beams) and also the beam when the second and fourth loudspeakers are reversed in polarity relative to the

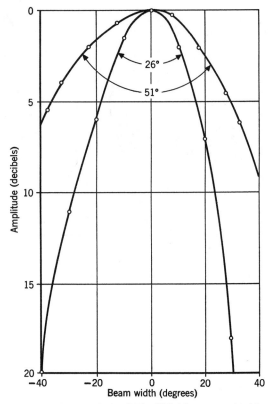

Figure VII-16. When the five speakers of Fig. VII-15 are energized in phase, a 51-degree beam (at the 3-decibel points) results. When alternate speakers are reversed in polarity, the measured beam width drops to 26 degrees.

other three. It is seen that a reduction in beam width does indeed occur, the extent of this decrease being exactly in accord with the predicted amount.

The visible patterns generated by the array for each of these two situations are shown in Figs. VII–17 and VII–18, which demonstrate how this effect comes about. When all speakers are in phase (Fig. VII–17), wavefront curvature becomes evident very early in the radiation pattern and purely linear wave fronts exist only at or very near the array itself. For the superdirective situation (Fig. VII–18) the linear wave fronts extend well along into the radiation pattern, and furthermore their length (for any significant sound-intensity value, i.e., for any brightness) becomes substantially greater than the length of the array (the five loudspeakers). A

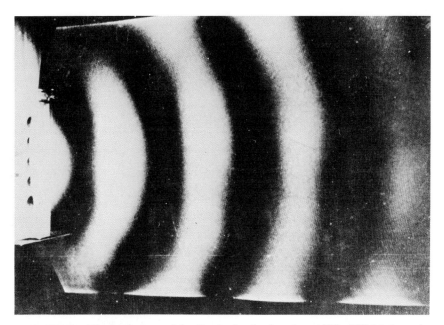

Figure VII-17. The in-phase condition for the five loudspeakers of Fig. VII-15 shows the expected curved wave fronts.

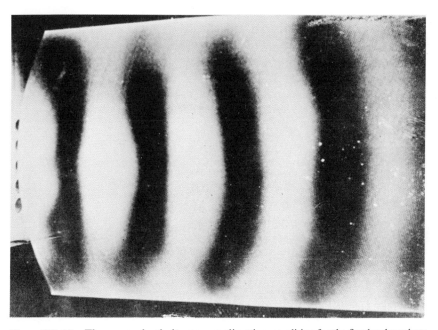

Figure VII-18. The reversed-polarity, or superdirective, condition for the five loudspeakers yields flatter wave fronts and hence a sharper beam pattern.

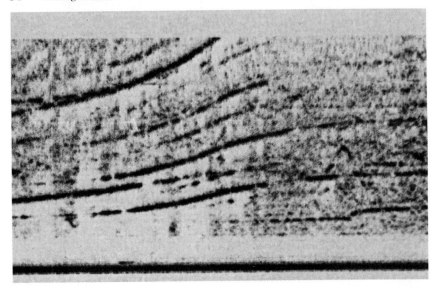

Figure VII-19. The spectrogram of the sound of an airplane passing overhead shows a doppler shift in its frequency components. (Time increases toward the *left* in this figure).

linear wave front like this, positioned well in front of the array, is thus equivalent to a wave front radiated by an array of a correspondingly greater length and hence generates a sharper beam than does the shorter, equally energized array.

Figure VII-20. Adding a delayed replica of the sound of a passing aircraft causes the recorded lines to separate during the downward glide.

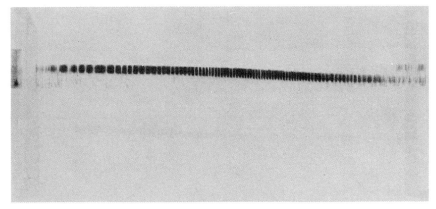

Figure VII-21. A broad-band analysis of the record of Fig. VII-20 discloses a beating of the two signals, the beat being most rapid at the middle of the downward glide.

Mechanical Sounds

The sound spectrograph offers a convenient means for examining the character of mechanical sounds generated by power plants of various sorts.

Figure VII–19 is a record of the sound of a propeller-driven aircraft as it passes overhead. The beginning and the end of the record show the numerous lines (frequencies) corresponding to the piston-engine vibrations and the propeller noise itself. Their shift in frequency during the course of the record is caused by the Doppler effect. As the plane approaches the listener (or the recording device), the pitch is raised by virtue of this motion; as it departs, the pitch is lowered.

An interesting result occurs in the case of this Doppler-modified frequency pattern when the signal picked up by the microphone is combined with the same signal delayed slightly in time. Figure VII–20 shows the narrow-band analysis of such a combination signal with one very strong frequency component. At the beginning and end of the signal nothing unusual is observed, but during the drop in frequency the undelayed signal leads the delayed one and the two lines separate.

Figure VII–21 is a broad-band analysis of the same record. Here the filter is so broad that the two lines are separated throughout the record. There is, however, a beating effect, with the beat frequency changing most rapidly at the center of the downward glide. This central point corresponds to the instant when the airplane is directly overhead.

Figure VII–22 is a record of a diesel engine, made by slowly increasing and decreasing the engine speed. Increasing the speed raises the lowest frequency, the fundamental, along with all of the harmonics (the higher lines). At the highest speed the harmonics exhibit the largest separation.

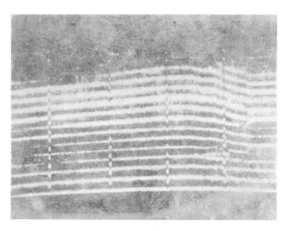

Figure VII-22. The spectrogram of the sound of a diesel engine shows it to consist of many harmonics.

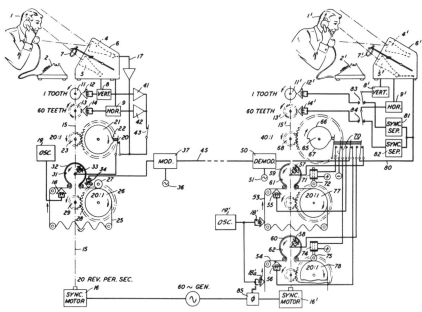

Figure VII-23. A picturephone system that permits the picture information to be transmitted over an ordinary telephone line. (From a patent issued to R. L. Miller on the author).

Figure VII-24. Viewing procedure for an experimental picturephone system based on Fig. VII-23. The insert shows the limited picture resolution available.

Picturephone Sounds

In one experimental version of a telephone link that includes the feature of the speakers' also being able to see one another the information content of the transmitted picture is greatly reduced and a new picture is transmitted only every 2 seconds. These steps permit the entire picture signal to be transmitted over a standard telephone (voice channel) line, achieving thereby almost the ultimate in economy for the user, but at the cost of a picture's having very low resolution.

The signals for these pictures, confined as they are to a voice channel, are in the audio range, and, if connected to a receiver, can be heard as sound. To make *these* sounds visible, however, techniques more like those employed in true television are used. One procedure involves recording the incoming signal magnetically for 2 seconds, then playing the recording back rapidly over and over again (at 30 times per second) and feeding the output signal to a cathode-ray tube. This procedure is sketched in Fig. VII–23. The viewer thus sees the one picture of the opposite

Figure VII-25.   Picturephone records made on electrosensitive paper.

party repeated 60 times (i.e., for 2 seconds), while the next incoming 2-second signal is being magnetically recorded on a second drum or tape loop for later replay and 2-second viewing. Figure VII–24 shows the method used in viewing the picture; the small insert is a photograph of a picture produced on a cathode-ray tube by this process.

A second procedure involves a sound-visualization method somewhat similar to that used in the Potter sound-spectrograph approach. During the 2 seconds in which the sound of the picture is being transmitted a stylus (or a rotating spiral wire) is caused to traverse many times a rectangular piece of paper. This paper is similar to that employed in the sound spectrograph in that it responds to electrical signals by becoming blackened when the signal is strong, gray when the signal is weak, and remaining white when the signal is very small. A "print" of the incoming signal is thus generated during the 2-second interval. It is presented to the viewer for 2 seconds while the next "sound-picture" is being printed. Examples of pictures produced by this procedure are shown in Fig. VII–25.

Loudspeaker Patterns

A knowledge of the sound patterns of loudspeakers can be helpful in ascertaining the accuracy with which the speaker has been assembled. Thus in Fig. VII–26 the pattern of the radiated sound is quite asymmetrical. Just above the main radiation lobe, a very strong hole appears in the pattern — a hole that is not present in the lower part of the pattern. The loudspeaker pattern in Fig. VII–27 is much more symmetrical, indicating a superior acoustic performance.

Figure VII-26. The sound pattern of this loudspeaker is asymmetrical, suggesting that the construction is faulty.

Figure VII-27. The sound pattern of this loudspeaker is quite symmetrical.

Index